REVOLUTION IN MINIATURE

REVOLUTION IN MINIATURE

The history and impact of semiconductor electronics

ERNEST BRAUN and STUART MACDONALD

CAMBRIDGE UNIVERSITY PRESS

CAMBRIDGE

LONDON · NEW YORK · MELBOURNE

Published by the Syndics of the Cambridge University Press
The Pitt Building, Trumpington Street, Cambridge CB2 1RP
Bentley House, 200 Euston Road, London NW1 2DB
32 East 57th Street, New York, NY 10022, USA
296 Beaconsfield Parade, Middle Park, Melbourne 3206, Australia

© Cambridge University Press 1978

First published 1978

Printed in Great Britain at The Pitman Press, Bath

Library of Congress cataloguing in publication data
Braun, Ernest, 1925–
Revolution in miniature
Bibliography: p. 201
Includes index
1. Electronics – History   2. Semiconductors – History
I. MacDonald, Stuart, joint author   II. Title
TK7809.B7   621.3815'2   77–82489
ISBN 0 521 21815 2

CONTENTS

PREFACE

Semiconductor electronics is one of the major technologies of the post-war period. Thousands of words have been written about it, yet there is not a single book available to the intelligent lay reader which tells the story in all its ramifications. Our aim in this volume is to present a fairly complete account of how the main inventions came about, how a major new industry grew up and how our lives have been affected by it all.

We have tried to avoid incomprehensible technical language and hope that readers with many different backgrounds will not only understand but actually enjoy the book.

We are deeply grateful to the many, many people who have so generously given of their time to enlighten us, and to those who have made available the results of their own research. In particular, we would like to thank Drs William Finan, Anthony Golding, Herbert Kleiman, Jerome Kraus and John Tilton. Without their help, this book would not have been possible. We are also grateful to the Social Science Research Council for a generous grant, which was vital to the completion of the task.

Many opinions have been expressed to us and much help given, but the responsibility for the content and accuracy of the whole text, including the opinions quoted, is entirely ours.

ERNEST BRAUN
STUART MACDONALD

Technology Policy Unit
University of Aston in Birmingham
March 1977

CONVERSION TABLES

£ Sterling – $ U.S. exchange rate

	£1 =
1947	4.03
1949	2.80
1967	2.40
1971	2.60 (U.S. devaluation)
1972 (September)	2.45
1975 (September)	2.35 } (floating £)
1977 (February)	1.71

U.S. wholesale price index
(1967 = 100)

1947	80
1951	90
1953	85
1958	90
1963	90
1967	100
1972	120
1973	140
1975	180

An innovation based on science

A technological innovation is like a river – its growth and development depending on its tributaries and on the conditions it encounters on its way. The tributaries to an innovation are inventions, technologies and scientific discoveries; the conditions are the vagaries of the market-place.

There are several special features to the innovation which we might call semiconductor electronics. First of all, it is not really a single innovation at all, but consists of a long series of linked innovations. There is a continuous chain of new electronic devices, based on semiconductor materials, which stretches from the cat's whisker used at the turn of the century to the latest and most astonishing achievements of modern microelectronics. The totality of these innovations, incorporated under the one name 'semiconductor electronics', is so massive as to make most single technological innovations pale into insignificance.

The scale of this innovation is matched by its impact. Whole new industries arose, new professions, new ways of production and new organisations. Without solid state electronics, we would have no powerful computers, no large-scale automation, no communication satellites or space exploration. There would be no electronic calculators or digital watches, no transistor radios, portable tape recorders or bugging devices. Many diagnostic methods in medicine depend on the new electronics, as does the heart pacemaker and the modern hearing aid. The impact of all these things on our lives is quite dramatic. For example, we can watch events in Australia while they are happening. The bank clerk no longer keeps accounts but only acts as an intermediary between the customer and the computer. Many goods are manufactured and assembled by automatic machines which resemble the science fiction robot in everything but shape. Air liners are guided by automatic systems and airline seats are reserved world-wide by an instantaneous computer booking system. The list could be extended almost indefinitely and is growing daily.

Apart from the scale of the innovation and the scale of its impact, the technology of semiconductor electronics is distinguished by its very great dependence on science. Perhaps more than any other innovation, modern electronics owes its existence to science; it is truly an innovation based on science.

Perhaps this feature does not surprise many observers. After all, we have come to accept the near synonymity of science and modern technology. But this view represents a gross over-simplification of the complex and manifold relationships between scientific discovery and technological development. Science is perhaps influenced by technology as much as technology by science. The view that science invents and technology develops and manufactures is quite untenable on close inspection, though it does come closest to the truth in the particular innovation we are concerned with – semiconductor electronics. Indeed it may be argued that the science-based invention of semiconductor electronics was one of the main factors in making this simple view of the relationship between science and technology so widely accepted. For obvious reasons this view is not unsupported by scientists and may have played a major role in shaping science policies over quite a long period. If pure science can bring about the introduction of major new technologies, then pure science deserves every support it can get on purely economic grounds. The support of science becomes expedient, rather than benevolent. Scientists would be the last people to complain about this public attitude.

> We can say that whatever the investment [which produced integrated circuits] was, it was well worth it. If we could think of other ways of investing Government money in some new industry, we would do it. It was well worth the investment.
>
> Quentin Kaiser

Times have become harder, and science policies have become less generous and more critical. The theory of technological innovation has also become more sophisticated, even if theory may still be a rather flattering term for a series of observations and conclusions, some of which are quite contradictory.

Some features of innovation have become reasonably clear, even if not universally agreed; others are more controversial. The main items of controversy are concerned with the role of science, the type of research required and the relationship between research and development.

The first requirement necessary to ensure that an invention reaches the stage of production and marketing, in other words becomes an innovation, is that a market for it should be believed to exist. We are specifically concerned with technological innovation and not that which occurs in the whole range of man's other activities. If we restrict the use of the word innovation to mean successful innovation, innovation that does indeed achieve at least some sales, we may say that a market must exist for the potential innovation. The mechanism by which the potential innovator assesses the market potential of his would-be innovation is known as market coupling. It may consist of market or opinion research, or it may consist of detailed negotiations with potential customers. The type of market coupling required depends on the type of innovation attempted.[1]

Leaving aside the actual mechanism by which market coupling is achieved, writers have distinguished between market-pull and invention-push.[2] These categories, by no means in water-tight compartments, are used to describe two possible ways of achieving a market for a new product: the innovator either perceives a need and produces a product to satisfy it, or the innovator feels such confidence in the appeal of his product that he believes it will create a demand for itself.

It may be argued that the first transistor falls into the market-pull category of innovations. There already existed a large market in electronic amplifiers and switches, and their obvious drawbacks were there for all to see. The valve was bulky, fragile, consumed a lot of power, had a restricted life and therefore restricted reliability, and required a fairly high voltage to operate. Any solid state amplifier, or indeed any amplifier, which could overcome all or some of these drawbacks, without introducing major new disadvantages of its own, would be a clear market winner. No need for elaborate market research, no need for sales campaigns to create a market; a good market for a good device was assured and was never questioned by anyone.[3] Indeed the creation of a solid state amplifier was more in the nature of fulfilment of an old dream than of a new invention.

The status of some of the other innovations in the long chain

leading from the early transistor to the modern integrated circuit
and its many new uses is more doubtful. How many were brought
about by market-pull and how many had to create their own
market? Can the distinction always be made and be meaningful?
Some examples stand out. Few would argue that a genuine need for
electronic pocket calculators existed prior to their introduction to
an unsuspecting market-place. Yet they have become a major commodity and few of us seem to be able to manage without them.
Perhaps the question of market coupling will become clearer as the
story of semiconductor electronics unfolds, for no device, no invention, however ingenious, can survive without a market.

The favourite figure in the repertoire of the theoretician of innovation is the 'product champion'.[4] This may be a person or a
group of persons, normally within the innovating organisation, who
are enthusiastically supporting the innovation and attempting to
push it through their organisation. The product champion may be
the inventor, or he may be the head of the section in which the invention was made, or he may simply be an enthusiast for the particular innovation. The motivation is immaterial; what matters is
that the product champion must be fully committed to the innovation and that his advocacy must carry weight within the organisation. The nearer the product champion is to the top of the tree, the
better the chance that he will convince the organisation to put
enough effort into the innovation.

The effort required may be very great and the financial risks enormous. It has been estimated that for every hundred pounds spent in
a research laboratory on an invention, at least a thousand pounds
must be spent in the development laboratory and pilot plants to get
the invention into a form in which it can be manufactured.[5] The
value of such estimates is doubtful because they assume the simple
mechanism by which a new product or process progresses from
research laboratory to development laboratory to pilot plant to
production. This model is grossly oversimplified in the general context of innovation, but for innovation in the semiconductor electronics industry it may apply as a first approximation. In any case,
simplified as the model might be, it illustrates correctly the fact that
inventing is incomparably cheaper than developing a new product
and launching it onto the market.

If success could be guaranteed by obeying all the rules of the
game, innovation would cease to be a game. A game it definitely is;

though whether it be a game of chance involving a lot of skill or a game of skill with a large element of chance is yet to be determined.

The initial innovation in semiconductor electronics, which heralded the era of solid state electronics and led directly to modern integrated circuitry with all its consequences, was the transistor. We may speak of the transistor as a single innovation and the rest of the development of solid state electronics as a series of innovations. It is the transistor which is the innovation based on science.

Inventing the transistor meant the fulfilment of a dream; almost the *perpetuum mobile*. Generations have dreamt of the *perpetuum mobile* and scientists, trying to give reality to the dream, concluded that the realisation was impossible. The impossibility of constructing a *perpetuum mobile* became the cornerstone of thermodynamics and inspired considerable progress in the field of heat engines and their theory.

In the case of the transistor, things turned out rather more happily. When the solid state amplifier was attempted time and again, science was always intimately involved. In fact, unlike many technologies, and particularly heat engines, the attempts at producing solid state amplifiers invariably originated in the research laboratory. Though attempt after attempt failed, it was clear that there was no basic conflict with nature involved. What was lacking was more knowledge, more understanding, better theory, better experiment.

Three things delayed the innovation of a solid state amplifier: the success of valve amplifiers, lack of knowledge of solid state physics and the lack of a product champion. The three factors are closely related. To succeed, a product champion had to convince the commercial managers that their product would not only work, but would work better than the valve. As most early inventors were pure scientists, they were usually only too well aware of the limitations of their devices. Those who were not were loners who could not raise sufficient investment and a sufficiently large team to come anywhere near a practical, competitive device.

By the end of the war the conditions for the innovation to succeed were at long last fulfilled. Enough had been learned in the thirties and during the war to be able to tackle the problem of producing reasonably pure germanium and to know with reasonable confidence what kind of electrical properties it would have and how to alter them. The limitations of the valve were also asserting themselves; anybody who ever laboured with a walkie-talkie knew that there must be better ways of going about transmitting and receiving signals. Finally, faith in the power of science was riding high. Science had played a dominant role in winning the war; science would play a dominant role in building a new and better world.

At Bell Telephone Laboratories all the factors came together.[6] Here there was no lack of product champions; the organisation was convinced that solid state switches and amplifiers or other components could play a major role in the future of telecommunications. The transistor itself, the solid state amplifier, had powerful protagonists in William Shockley, Walter Brattain, John Bardeen and many others. The market for such devices was never in any doubt; if they could be made, they could be sold. If nobody else bought them, certainly the vast Bell empire itself would form an adequate market. Neither was capital a problem in such a vast organisation as Bell Laboratories. But more important than money were human scientific resources – brain-power – and that again was a commodity in rich supply at Bell, where there was a vast concentration of first-rate scientists, backed by the best available facilities in instrumentation and technicians.

With hindsight, the innovation under these circumstances could not have failed. Certainly the managers at Bell Laboratories did not lack the foresight to attempt the fulfilment of the dream.

Theoreticians often classify innovation according to the market aimed for. They speak of innovation in large and small capital goods and of innovation in consumer goods. Large capital goods normally have only one potential customer and that is usually a public body or large corporation. Small capital goods have a larger number of potential customers, but they are still experts in their field. In both these cases, continuous liaison between innovator and market is essential; in some ways the customer shapes the innovation and adapts it to his needs.[7] In consumer goods the situation is different. It is, of course, possible to conduct market research and do pilot marketing studies, but essentially the innovator is left to guess the consumer's preferences. This is especially so in the 'invention-push' situation, but holds to some extent in the 'market-pull' case. Although a requirement may be known, it can still vary sufficiently in shape and size and form to make all the difference between success and failure.

Semiconductor innovation really falls into all these categories. By its very size and pervasiveness, this innovation eludes most traditional classifications. This is not surprising if we consider the whole family of innovations of which modern semiconductor electronics is comprised. But even if we only consider the ancestral innovation of the family, the transistor, we would be hard pushed to accord it a proper place in the classification. Undoubtedly transistors are not consumed directly, but they are bought by innumerable amateur radio and hi-fi enthusiasts to build into their sets. Nevertheless, perhaps the classification as small capital goods is more appropriate as most transistors find their way to manufacturers of electronic equipment. The 'standard customer' is an engineer who will build the transistor into his company's equipment. The ultimate user may, of course, again be the man in the street buying a television set, or it may be the Military, or the Post Office or any firm using electronic control.

Much has been said about innovation, but no mention has been made of the most fundamental question of all; why innovate? Unfortunately, most fundamental questions have no easy answer. Our question is no exception. Nevertheless, elements of a complete answer can be obtained and these may help to illuminate the question. Some elements of the answer appear to be: the desire of the individual for new things and for exploring new possibilities; the stimulus innovation gives to economic growth and therefore increased material well-being; the necessity to innovate in order to remain competitive.

There is no doubt that successful innovation is a major factor in economic growth; it is the very stuff of growth. Unsuccessful attempts to innovate, on the other hand, waste resources and retard growth. Even successful innovation can waste resources if its pace is too fast or too slow. A low rate of failure in attempted innovation and the correct choice of pace are the elusive secrets of economic success in the modern world.[8] The story of semiconductor electronics, as it unfolds, will provide examples of all these things. Firms trying to innovate too fast go under, firms that stay behind in the race fare no better. The totality of semiconductor electronics becomes a vast new industry and a major contribution to post-war economic growth. But above all, individual scientists are spurred on by their desire to know, to beat Nature at her own game, to use their knowledge to produce something no-one else has thought of, to do things no-one else has achieved.

The uniqueness of the semiconductor innovation does not lie in the desire to produce new things for the spirit of invention is as old as mankind. The uniqueness lies in the fact that nobody other than a highly qualified scientist could have made any contribution in this field. There is no way in which semiconductor devices can be tinkered with at home; no way by which a skilled craftsman could improve their performance. The mode of operation of these devices is so complex and intricate, the scale so small, the interactions so subtle, that all old-fashioned inventiveness proved of no avail.

Perhaps scale is the most important factor. Everything in a solid happens on such a minute scale that not even a microscope, even an electron microscope, can resolve the elementary processes. The scientist must operate at a level of abstraction of which the untrained mind is not capable in order to visualise processes which cannot be seen. The same applies to purity of materials. Again, the level of sophistication is beyond the means of not only the layman, but of any single scientist. No more than one foreign atom in ten million correct atoms can be tolerated in the materials used for the manufacture of solid state electronic devices. The demands on purity and cleanliness exceed by far those made by the surgeon, yet even he cannot function without extensive scientific training.

Many innovations emerge from technology itself. They consist of novel ways of using technology, new twists to old experience. Even these innovations use science; to test ideas, to measure performance, to improve materials, to avoid pitfalls. But semiconductor electronics had no previous technology to fall back on and past experience was irrelevant. The only way it could emerge was by means of the deliberate use of a vast armoury of scientific knowledge.

Much controversy in the field of innovation centres round the topic of research and development, R & D in the jargon of the trade. Within the field, definitions such as pure and applied research, goal-oriented and free research and fundamental research abound, and are hotly disputed.[9] There is much heart-searching about the funding of all these. How much money should be spent on each category? Who should pay and how should the expenditure be controlled? Ammunition for the various points of view is sought from past experience, particularly from the eminently successful experience of semiconductor research. However, it is difficult to draw conclusions from this experience because of its multi-faceted nature. As with a multi-coloured garment of intricate design, the overall impression is what really matters. Little insight is gained by following individual strands, yet the strands make the pattern.

We shall return to these questions later, but for the moment several things stand out. There is no doubt that semiconductor electronics gained enormously, in fact was founded on, fundamental undirected academic research. All advocates of free research find strength in this example, where pure knowledge, acquired for its own sake, has led to a major new economic activity. There is no denying the historic fact, but whether this was the only path that could have been taken is an entirely different matter.

By the end of the Second World War, pure research had provided sufficient information to make a goal-oriented approach worth while. The approach at Bell Laboratories was still fundamental in that they did not empirically tinker with devices, but tried to understand at each step what was going on. But their approach was goal-oriented: they knew what they wanted and went looking for it. At this stage of the development it was vital that the research should be carried out by a large multi-disciplinary team of scientists, backed by a strong and financially powerful organisation.

> Always Bell was pursuing semiconductor research from the hope of a communications application. It was not in the department of non-applicable research ... Those guys were looking, hoping for new effects and they found not an effect; they found an amplifier which was to go beyond their dreams ... They were trying to find out what was going on and in the process of trying to find out what was going on they found an amplifier.
>
> James M. Early, formerly of Bell Laboratories

Once the first devices had come on the market, several new factors entered. One was the Military. There is controversy about its role in detail, but little doubt that by various means it put a lot of money into the development of semiconductor electronics and that this money, at the very least, accelerated the development. Another factor which entered at this stage, and became increasingly important, was the small entrepreneur with a limited range of scientific skills. Once the major research and development were done, the flexibility of small size became an advantage over the great power of the large corporation.

It seems that the development went full cycle; from the small academic laboratory where basic knowledge was acquired to the mammoth organisation where the knowledge was marshalled to yield an industrial product. And from the large industrial organisation the cycle turned again to favour the skilful entrepreneur. This time he was not a fundamental scientist, but a man of science willing to use a limited range of knowledge. Whether the cycle has now gone one step further, towards concentration of industrial power, is still questionable, but it appears likely.

Whether one wishes to advocate support for free research or for goal-oriented research, private enterprise or state intervention, large scale or small scale, the story of the seminconductor is large enough and varied enough to provide arguments for all sides. Only one thing remains beyond argument. No invention or innovation has ever owed more to pure, abstract science.

CHAPTER 2

Genesis

The main impetus for the development of solid state electronics came from three directions: the invention and development of wireless telegraphy and all that it led to in radio and television, the success of valve electronics, which widened the scope of electronics far beyond the original wireless telegraphy applications, and finally pure research in solid state physics which had, from time to time, almost by accident, brought about various solid state devices. We shall follow these three main strands of the development and the many side branches and interconnections. As always, some simplification will be necessary for it would not otherwise be possible to trace reality in all its infinite variety and complexity.

The story of wireless telegraphy can begin with Michael Faraday and his discovery of electromagnetic induction. In 1831, this extraordinary man showed that the motion of a magnet could induce the flow of electric current in a conductor in the vicinity of the moving magnet. The discovery was very rapidly taken up by numerous investigators.[1]

On the practical side, it was Weber and Gauss in Göttingen who had demonstrated by 1833 that the new effect could be used to pass messages from one end of town to the other. They used a magnet and a coil of wire and moved the coil in the magnetic field with the aid of a morse key. The signal produced in the coil was transmitted through a pair of wires strung across town and supported by a church spire, and was detected at the far end by a magnetic needle, rather like that in a compass, suspended in the centre of a coil through which the signal was fed. In this way the two eminent men communicated between their respective laboratories for years.[2]

The theoretical interpretation of Faraday's discoveries was given by Clerk Maxwell, when he published his famous Maxwell equations in 1865.[3] Maxwell's is one of the outstanding examples of a successful scientific theory. It still stands unchallenged and has served as the basis for thousands of further scientific publications which have developed the theory to cope with innumerable variations in the detail of the experiment. Every student of physics spends a good proportion of his course learning Maxwell's equations and their consequences. Their main strength lies in the fact that they do not describe a physical model; they simply give the mathematical rules which measured electromagnetic phenomena obey. These equations are one of the great classical achievements in the tradition of describing physical phenomena by mathematical equations, abandoning, perforce, the need to describe them in visual or imaginable pictures.

Maxwell himself could not foresee all the consequences of his equations. But, again in the best tradition of scientific discovery, he realised that the equations, describing phenomena already discovered by experiment, could be used to describe new phenomena yet to be discovered. The discovery of such predicted phenomena would serve as a touchstone for the 'veracity' of the theory. Successful prediction is taken as a kind of proof of the correctness of a theory, although strictly speaking, only the reverse should apply, i.e. unsuccessful prediction can be taken as a falsification of a theory.

One of the major predictions of Maxwell was the existence of electromagnetic waves. This means that if an electromagnetic signal is produced somewhere, by the motion of electric charges, such as electrons moving backward and forward (oscillating) in a wire, then this signal will propagate from its point of origin much as a wave is propagated in water from the point at which it is hit by a stone. The electromagnetic wave predicted by Maxwell is a mathematical description. It does not require that the wave propagate through any particular medium. Indeed, propagation can be through vacuum. It does give an accurate description of what will be the distribution in space and time of an electromagnetic signal.

At this point the story moves again from Britain to another small University town in Germany, this time Bonn. In 1888 Heinrich Hertz demonstrated experimentally that the electromagnetic wave, predicted by Maxwell, did indeed exist. An electromagnetic signal moves from its point of origin as a wave and can be detected at a distance. The scientific foundations for a new era in communications had been laid.[4]

In order to communicate, we must transmit electromagnetic

signals containing information. A continuous electromagnetic wave of constant frequency can be detected, but contains no information other than the frequency and amplitude (or wavelength and intensity, which amount to the same thing). If we wish to transmit more information, such as words or figures, we must transmit a sequence of short signals, perhaps in Morse code. For any practical system of communications to operate, a further condition must be fulfilled. Each receiver must be tuned to receive just the frequency of the one transmitter it wishes to receive, otherwise it will not be able to disentangle the required signal from any other signals which happen to be around. Therefore, the transmitting signal must be broadcast at a constant and pre-arranged frequency. These requirements lead us to the concept of the carrier wave, a more or less continuous electromagnetic wave of constant frequency used to tune the receiving station to the required transmitting station. The carrier wave carries no information beyond the identification of the transmitter by its frequency. The actual information is carried on this wave by periodic modulation of either its amplitude or its frequency. If we wish to transmit a Morse-coded signal, we increase the amplitude of the carrier wave for the time intervals prescribed by the message. The receiving station must be able to detect this modulation if it is to read the message borne on the carrier wave. For this purpose we use the properties of a rectifier, a device which allows current to pass freely in one direction but not in the other. Used in a circuit, a rectifier will convert an alternating current (a.c.) into direct current (d.c.). In a detector circuit, the carrier frequency is effectively rectified to give a d.c. voltage which is not processed any further by the receiver. The low frequency modulating voltage, on the other hand, is only inefficiently rectified and passes through the circuit little changed from the original a.c. modulating signal. We have thereby achieved the desired demodulation: instead of a carrier wave with a modulation superimposed upon it, we have the desired a.c. signal alone. The problem of detection of a signal is simply one of rectification.[5] The resultant a.c. signal can then either be fed directly to earphones or can be amplified to feed into a loudspeaker. This latter possibility presupposes, of course, that a means of amplification is available, and this development came long after the first transmission and detection of radio signals. Indeed even the method of detection by a rectifier was not the first method used.

The very first wireless transmissions were detected by a bizarre device known as a coherer. In some ways this may be regarded as the first solid state electronic device, although it is perhaps stretching the meaning of the term a little. The coherer consisted of a glass tube partially filled with metal filings. The tube had metal contacts on each end and formed part of a circuit with a battery and some kind of recording instrument. Normally, the metal filings were too loosely packed in the tube to conduct current and none would flow through the circuit. When an electromagnetic signal reached the coherer, its action caused the metal filings to coalesce – cohere – and thereby rendered the tube conducting. A current would flow through the circuit and actuate a buzzer or a recording meter. When the signal stopped, the coherer did not return to its initial non-conducting state by itself. The tube had to be tapped in order to loosen the metal filings again. In the most ingenious coherers this tapping was done by a little hammer actuated by the recording device. A signal was received, recorded and the coherer tapped and made ready for the next signal, all in one operation.[6,7] The success of the coherer was short-lived. Ingenious as it was, it was an inherently clumsy device which was difficult to develop to a degree of perfection required for routine operations on a large scale.

The first true detector rectifier was also the first true solid state electronic device, affectionately known as the cat's whisker of crystal radio fame. In 1874 Ferdinand Braun, a professor of physics at Marburg, discovered that the contact between a metal wire and the mineral galena (lead sulphide) was rectifying – it would conduct electricity in one direction only. A short time after this discovery, Braun also discovered or, more accurately, invented, the tuned electrical circuit. This makes it possible for radio waves of a single frequency to be received, while all other frequencies are rejected. With the tuned circuit, all the prerequisites for a wireless communications system were assembled. Carrier waves of a desired frequency could be transmitted and the receiving circuit could receive them.[8]

The cat's whisker rectifier flourished for a short time as the sole useful detector of wireless signals.[9,10] It was small, cheap, simple and good fun, but its reliability left a lot to be desired. It does not appear that these devices were ever manufactured on a large scale to rigid specifications. Every wireless manufacturer and wireless amateur produced his own cat's whiskers. All that was required was a suitable piece of galena, an insulating holder and a piece of thin, springy wire. The sensitivity of detection was critically dependent on the precise location of the wire on the crystal – an adjustment achieved by trial and error. The device was not exactly robust, but if the adjustment slipped, the matter was easily put right.[11]

Both commercially and technically, the cat's whisker devices were highly vulnerable. They had something of the delicate toy about them, which delighted the amateur but frustrated the commercial manufacturer and operator. Indeed, the cat's whisker was soon relegated to the exclusive domain of the radio ham, while the professional scene was taken over by a device which arose from an entirely different area of science.

The cat's whisker was a product of early solid state physics. It owed its existence to the fact that physicists like Braun occasionally investigated the electrical properties of curious substances, such as galena. These activities had been going on, on a small scale, since the discovery by Faraday in 1833, that the electrical resistance of silver sulphide decreased with increasing temperature while that of other conductors increased.[12] The other conductors were metals, while silver sulphide was the first semiconductor to be investigated. It took something like another hundred years before the term semiconductor came to be used and to acquire any precise meaning. In fact, it was not until the 1930s that solid state physics acquired any systematic, structured and theoretically supported framework of knowledge. The cat's whisker itself did not receive any theoretical explanation of its action until this period, although by then it had long been in a state of technological dormancy.

The device which caused the decline of the cat's whisker was the vacuum diode. Its two parents were the technology of manufacture of light bulbs and the discovery that electric current could flow from an incandescent filament into a second electrode. The light bulb had required two skills to make it possible; the ability to make electrical contacts through glass and to evacuate all air from the bulb. Vacuum technique became established with the brilliant demonstration of the mechanical vacuum pump by Otto von Guericke in Magdeburg in 1657 – one of the great combinations of showmanship and science.[13]

Once the manufacture of light bulbs was established – and by the beginning of the twentieth century a number of manufacturers were producing them – the step to the vacuum diode was only small. By 1904 Fleming in England had added a second, cold electrode to something like a light bulb to obtain the first vacuum diode.[14] Electrons can travel from the hot filament, the cathode, to the cold electrode, the anode. They cannot travel the other way, because the anode does not emit electrons into the surrounding vacuum. Therefore the device rectifies. One virtue of the vacuum diode was that the light bulb industry already existed to take up its manufacture. Other virtues were its relative reliability and the ability to handle quite large currents, at least compared with the cat's whisker. Its main disadvantages were its great bulk and the fact that it needed a good deal of electric power to feed its incandescent filament.

It seems unlikely that the diode alone would have captured the wireless market. What made it really commercially viable and sounded the death knell for the cat's whisker as a commercial device was the invention of the triode by De Forest in 1906.[15] The triode, as the name suggests, contains three electrodes in an evacuated bulb. The third electrode, called the grid, has the task of influencing the current which flows between the cathode and the anode. A small electrical signal applied to the grid can produce large changes in the current between cathode and anode and this forms the basis of an electronic amplifier. The fact that a 'vacuum valve', or 'vacuum tube' in America, can amplify electrical signals and that it can be readily manufactured, made this type of device a natural winner. Although it is possible to feed earphones directly from a cat's whisker detector and it is possible to operate slow, clumsy recording devices by coherers; it was only the availability of amplifiers which made possible the widespread use of radio and all that is associated with it. Once it was decided that amplification was worth buying, even at the price of using bulky vacuum valves with power-consuming incandescent filaments, it made sense to replace the cat's whisker with vacuum diodes. If an amplifier uses several triodes and needs a power supply to feed their filaments, then the addition of a diode makes little difference to either bulk or power consumption.

So, very shortly after the first experiments with wireless telegraphy, vacuum tube electronics had come onto the scene and determined the development of all electronics for the following fifty years. The triode was followed by vacuum tubes with four, five and even more grids. The tubes became smaller, more reliable, cheaper. The accompanying circuitry achieved a similar degree of sophistication. By the beginning of the Second World War broadcasting was commonplace, the penetration of radio receivers in the western world widespread, radio telegraphy universal and electronic devices used in a wide range of applications, from long distance telephone repeaters to burglar alarms. Even television broadcasts started just before the war and electronic instruments were in common use in scientific and medical applications.

At the beginning of the war, a very special requirement brought about a curious re-emergence of the discarded cat's whisker. It was found that extremely high-frequency electromagnetic waves could be used for the detection of distant metallic objects, such as aircraft. These high-frequency beams were reflected from the objects and the returning signal could be detected. The system, known as Radar, could locate objects accurately over a long range and thereby provide valuable warning of the approach of enemy aircraft or ships. But to detect high-frequency signals, the detector had to be of extremely low capacitance. Vacuum diodes of sufficiently low capacitance cannot be made and therefore the solid state, or cat's whisker, rectifier was resurrected.[16] In its wartime form it consisted of a small pellet of silicon instead of the old galena, and the metallic contact consisted of a thin tungsten wire. Much empirical development went into their manufacture, as the fundamental scientific knowledge of their mode of operation and the factors affecting their performance was still rather scant. What knowledge there was, was based on the emerging academic discipline of solid state physics, which had barely existed at the end of the First World War and had not reached maturity by the beginning of the Second.

In order to understand the scientific basis of modern semiconductor devices, we must look in a little detail at the period from 1920 to 1940. Atomic and nuclear physics undoubtedly stole the scientific limelight, but on a more modest scale parallel developments took place in the understanding of the physical behaviour of solids. The aspect of solid state physics of most direct interest to the historian of solid state electronic devices, is that dealing with the behaviour of solids in electrical and magnetic fields. Mechanical, thermal and optical properties are important, but what really matters is the electrical behaviour of solids.

By the early twenties it was generally accepted that metals conducted electricity well because they contained plenty of almost free electrons. The electrons in a metal are not free to leave it under ordinary circumstances, but they can move around within the metal rather like the molecules of a gas move within a jar containing it.

When an electric field is applied to the metal, more electrons will move in the direction of the field than in any other direction and a net flow of current will be established. A metal is, therefore, a conductor of electricity. This free electron theory of metals also explains the fact that metals conduct heat well. If one end of the metal is hotter than the other, electrons at the hot end will acquire additional thermal energy, which means that their random motion will, on average, become faster. The increased thermal motion will spread and conduct heat through the metal because the electrons collide with the atoms constituting the metal.

Metals vary tremendously in their properties. All are good conductors of heat and electricity, but some are a lot better than others. The resistivity of different metals can differ by orders of magnitude. That of pure silver at room temperature is almost a hundred times smaller than that of mercury. Melting points, hardness and strengths also differ by very large factors. Probably as a result of these very large variations within the metal category of solids, substances which do not properly belong to this class seemed to fit in quite readily. The existence of a class of materials now known as semiconductors was not immediately obvious because most semiconductors fitted quite neatly into the category of metals. Those that did not could be classed as insulators, materials without free electrons and therefore poor conductors of electricity or heat. The tolerance for variations within the classes of metals and insulators had to be great and the special classification of semiconductors did not suggest itself to investigators for a considerable time.

A number of curious phenomena had been observed over the years, but it often takes more than a series of isolated observations to establish a new field and to overthrow accepted beliefs. A measure of critical doubt is necessary for progress, yet excessive doubt paralyses the will and is counterproductive. The scientific community regards the occasional freak experiment as something to be set aside for the moment, to be looked at later or in conjunction with other results which might help to make some sense of it.

Amongst the curious phenomena randomly observed and now accepted as characteristic properties of semiconductors, it is customary to list four. In 1833 Faraday discovered that some conductors, notably silver sulphide, became better conductors as their temperature rose. Most conductors, indeed all true metals, become worse conductors with an increase in temperature. In 1839

Becquerel noticed that a voltage appeared at the junction between a conductor – what would now be called a semiconductor – and an electrolyte, when this junction was illuminated. We now call such an effect a photo-voltaic effect and it occurs only at semiconductor junctions. The change of conductivity of some substances under illumination, now called photoconductivity, was first discovered by W. Smith, working on selenium in 1873. The discovery by Ferdinand Braun in 1874 of the rectifying properties of contacts between metals and semiconductors completes the traditional list.[17]

Perhaps we should add two more phenomena which were observed in the late nineteenth century and were to play a decisive role in semiconductor research. W. Thomson was the first to observe changes in the resistance of some materials when a magnetic field was applied to them, and E. H. Hall found that a voltage could be established between opposite faces of a slab of material in a magnetic field. The Hall effect is crucial in establishing whether conduction in a material occurs by the normal transport of electrons or by a very special mechanism which can be described as transport by positive particles, known as holes. We shall return to this concept later.[18]

Only two other discoveries of importance to semiconductors were made before the First World War. The event of major importance was the postulation of the quantum hypothesis by Max Planck in 1900. According to this hypothesis, light waves interact with matter as if they were made up of packets of energy, called quanta. Each quantum of energy may in some ways be regarded as a particle of light, a photon.[19] Quantum theory eventually had the most far-reaching consequences for the whole of physics. Indeed, it precipitated a major scientific revolution which profoundly influenced thinking in all areas of science, and semiconductors were no exception. This major development, however, had to wait until the twenties. The fact of immediate importance to the emerging physics of solids, and we still mean metals and insulators with semiconductors included within these categories, was the application by Einstein of the quantum hypothesis to the photoelectric effect.

This is a classical example of one of the major ways in which new discoveries are made and new ideas conceived. It might be called external fertilisation, because an idea originally developed for the solution of one problem, often proves fertile in solving an entirely unrelated problem in a different sphere of science. It stresses one of the characteristics and one of the fascinations of science; the discovery of links between apparently unrelated phenomena. When connections and interrelations are found, factual observations can be built into the edifice of science.

For a long time light had been described as a short wavelength (high-frequency) electromagnetic wave. However, when light interacted with matter to cause the emission of electrons, many puzzling features occurred which the wave theory could not explain. Why was only light above a threshold frequency effective in causing the emission of electrons when it was shone on a metallic surface? Why was the number of emitted electrons proportional to the intensity of illumination, but their energy proportional to the frequency of the light? The energy of a wave is proportional to intensity and should have had nothing to do with frequency. The puzzle was resolved when Einstein adopted Planck's quantum hypothesis to explain the photoelectric effect. If light consists of small packets of energy and their energy is proportional to the frequency of the light, then all is explained. Indeed, the success of Einstein's explanation was so overwhelming that very few people opposed the revolutionary idea. Planck had originally proposed the hypothesis for the solution of an entirely different dilemma, the so-called ultraviolet catastrophe in the classical theory of thermal radiation.[20]

The photoelectric effect, as explained by Einstein in 1905, proved of crucial importance to solid state physics and to the eventual emergence of the transistor. The first effect of the new theory was to give a new fillip to experimental research on the emission of electrons from solids, in particular the release of electrons by light. This photoelectric effect was studied mainly, but not exclusively, on metallic surfaces. The surface to be studied had to be kept in a good vacuum, because otherwise the electrons emerging from it could not be collected. In air, electrons cannot travel any distance because of collisions with gas molecules.

One of the research schools in which the photoelectric effect was investigated was in Berlin and a young physicist, R. W. Pohl, was amongst the people working on the problem. This involved the preparation of clean metallic surfaces, often by evaporation in vacuum, and careful measurements of emitted currents as the wavelength and intensity of illumination were changed. The required vacuum was obtained by mercury diffusion pumps and these required a supply of liquid air to condense the mercury vapour. The First World War interrupted the work in Berlin as elsewhere and when Pohl returned to peace-time physics, now as a professor in Göttingen, the impoverished post-war economy could not supply him with liquid air. Work on the external photoelectric effect was therefore ruled out. But if light causes electrons to be emitted from solids, then surely light must also cause changes in electron energies inside a solid. And if vacuum was required to study the emission of electrons and vacuum could not be attained in post-war Germany, then one had to turn one's attention to effects inside the solid, which occurred independently of the external surroundings. By such reasoning, Pohl was driven to the study of the photoelectric and luminescence properties of solids.[21] Other considerations and influences played a role in this decision, which we are recalling here not only because it proved of great importance to the foundation of semiconductor physics, but also because it is characteristic of the tortuous way in which real history, scientific or otherwise, is made.

Luminescence, the emission of light by some solids when illuminated by radiation of suitable energy, was a phenomenon well known by the early twenties, though not properly understood in the present sense. However, W. C. Roentgen, the celebrated discoverer of X-rays, had studied the phenomenon extensively and Pohl had always been interested in Roentgen's work. When Roentgen published a lengthy paper on the subject, too long for most people to read, Pohl studied it carefully and found many puzzling features in it. The scene was set: there were puzzles to solve, there was an incentive to work on internal effects in solids, Pohl had previous experience on the photoelectric effect and an incentive to use his knowledge in a new field.

Work started in Göttingen in the early twenties. The first solid to be investigated was zinc sulphide in powder form, a substance extensively used as a luminescent material. A correlation was established between a change in conductivity and the emission of luminescence when the powder was illuminated. The choice of material had been made for its luminescent properties, but at the same time Pohl realised that if he wished to study changes of conductivity with illumination, the internal photoelectric effect, he had to choose a substance which was a poor conductor in the dark.

By a lucky coincidence, though perhaps not all that lucky in retrospect, Pohl was given a diamond crystal to experiment with. He was pleased to get away from work on a powdered substance because only thus would it be possible to be sure how electrons behaved in a crystal. The plan was to use the Hall effect to measure how fast the electrons liberated by light could drift in an electric field. Before electrons in an insulator can travel at all, they have to be liberated from their normally localised states (in which they are tightly bound to their parent atoms) into energy states in which they are free to travel in the crystal if an electric field is applied to it. The diamond was clamped inside the gap of an electromagnet and all was ready. The current was switched on, the powerful magnet seized the clamps holding the crystal, and the precious stone shattered into a thousand pieces. By an unfortunate oversight the clamps had been made of steel. Pohl swore never to try Hall effect measurements again, a decision he regretted to his dying day.[22]

Nevertheless the appetites of Pohl and his colleagues had been whetted to use single crystals in future experiments. They turned to natural crystals of rock-salt and, more precious than gold or even diamond, a natural crystal of potassium chloride. Only several years later did Pohl's group develop the art of growing artificial crystals sufficiently to produce an adequate supply of experimental materials. Painstakingly, over years of careful experimentation, did Pohl and Hilsch and others build up concepts which form the basis of modern semiconductor physics.

The picture that had emerged by the early thirties was of electrons in an insulator normally tightly bound to their parent atoms. They could, however, be freed by light and could then travel some distance if an electric field was applied. How far they can travel depends on the type and quality of the crystal. Generally, it is not very far because they become trapped in a variety of localised imperfections in the crystal. In some crystals, particularly crystals of the rock-salt type, photoconductivity is only possible if special centres have been created. These can be referred to as either colour centres, or as donor centres because of their ability to donate electrons to the conduction process. Thus, by the early thirties, many of the modern concepts had been clearly expressed. The path of an electron in an insulator is limited by various imperfections and the availability of electrons for conduction is influenced by imperfections known as donor centres generally consisting of small amounts of the right type of impurity. It was also clear how luminescence, photoconductivity and the colouring of crystals related to each other and to the optical absorption spectrum of the crystals.[23,24,25]

Curiously, some older concepts lingered on for a while. Referring to photoconductivity in selenium, Gudden and Pohl wrote in 1925 'We incline to the opinion that the resistance of the conduction path is reduced by a coherer-type bridging by the photoelectric effect of the many boundary layers present even in macroscopically uniform crystals . . .'. The concept of grain boundaries and imperfections was modern, but the comparison of grain boundaries with the powder in a coherer seems quaint.[26]

In 1933, on the occasion of a speech at a ceremony to mark the 100th anniversary of the Weber and Gauss telegraph, Pohl showed that electrons could enter one end of a crystal, such as rock-salt, from a pointed metal electrode and travel for a considerable distance towards the plane electrode at the other end. The beauty of the experiment is that the migration of electrons can be visually observed; they travel as a cloud of a vivid colour in an otherwise colourless crystal. This is not caused by any inherent colour of the electrons themselves, but by the formation of colour centres during the movement of electrons through the crystal.

In that same speech, Pohl said that, in his opinion, vacuum tubes in radios would eventually be replaced by small crystals in which the motion of electrons could be controlled.[27] Indeed, in 1938 Hilsch and Pohl published a paper in which they described a crystal analogy to the vacuum triode. They used a crystal of potassium bromide and caused electrons to enter it from a pointed metal electrode. In addition to the plane anode at the other end of the crystal, a wire was inserted into the crystal near the cathode, rather like the grid of a vacuum triode. By changing the voltage on the control wire, the current flow to the anode could be controlled and amplification obtained. The device had no immediate practical use; it was too slow and worked only when the crystal was heated. But it demonstrated that solid crystals could serve as amplifiers in electronic circuits and that was all Hilsch and Pohl wished to show.[28]

> We had no practical aims in mind . . . Either one works in a University, or one is a man who gets involved in litigation about technical devices. R. W. Pohl

For Pohl there never was any doubt that solid state amplifiers would come about, but his task was the understanding of the physics of crystals, not the design of practical devices. Indeed, Pohl was a little suspicious of any kind of design, even the design of theories.

> Theories come and go; experimental facts are there to stay.
>
> R. W. Pohl

Despite the fact that most of the relatively few physicists worked on nuclear and atomic physics during the inter-war years, further important developments in solid state physics took place during those years. Again, much of the most important progress was made by the principle of external fertilisation. Theories originally developed to explain phenomena in atomic and nuclear physics were applied to explain solid state phenomena. The major theory of the day was, of course, quantum mechanics, arising principally from Planck's hypothesis and the well-known equations treating wave phenomena. Bringing these together, because light in some ways behaves as a wave and in some ways as particles (photons), Schrödinger produced his famous equation. Again the bringing together of ideas from unrelated fields yielded a new idea, and this time it yielded the foundation of a major theory which is still current and still fruitful.[29]

Once the Schrödinger equation was published, an explosion of activity in physics occurred. As if a barrier had been opened, there was a sense of surging forward. Quantum mechanics developed rapidly and a number of bright young men, who otherwise might have remained obscure, gained well-deserved fame in developing and applying the new ideas.

The most important consequences for solids were realised within the next six or seven years by people like Sommerfeld, Bloch, Wilson and many others. It was realised first that electrons in a metal behave like a very special kind of gas, its main peculiarity being that no two particles can have exactly the same energy. Electrons also have wave-like properties. Just as a column of air in an organ-pipe will vibrate only at certain frequencies, so only certain frequencies are permitted to electron waves in a solid. But, as we have seen before, the frequency of the wave determines the energy of the particle and we therefore have a situation in which some energies are permitted to electrons and others are not. Electrons face two restrictions on their energies: some energies are altogether forbidden to any electron in the solid and also any energy state occupied by one electron is forbidden to all the others. It is a little like a bus on which no passenger must be standing and any seat already occupied is un-available. The result is that if a bus is full, the next hopeful passenger in the queue must wait for the next bus. Energy states in a solid are arranged in bands; when one band is full, further electrons must go on to the next. Between bands there are forbidden gaps and electrons cannot have an energy which would fall into such gaps. This view of electrons in solids is called the band theory of solids.

When all the electrons in the solid have been neatly fitted into their energy bands, rather like a works outing into the available coaches, two situations are possible. The uppermost energy band containing electrons may be completely full or it may be only partly filled. If this band is only part-filled, then there are energy states available to electrons which are just above the highest filled states. When we apply an electric field to the solid, the field will exert a force on the electrons and will accelerate them in the direction of this force. The electrons thus affected will gain some additional velocity and, therefore, energy. In band theory parlance we say that they move up slightly within their band. The net result is, of course, that we get a net increase of velocity in the direction of the field and we observe the flow of an electric current. The material is a conduc-tor of electricity and, therefore, a metal.

Now take the other possible case; that the last filled band is com-pletely full. If we apply an electric field to such a material, we observe no conductivity and we class the material as an insulator. The electric field will still exert a force on the electrons, but it cannot accelerate them as they are not allowed to gain a small amount of additional energy. There are no higher states of energy within the band available to them.[30]

The band theory neatly classifies materials into metals and in-sulators, in good agreement with observation. But what about the class of semiconductors and what about all the experimental facts about photoconductivity?

Essentially three things can make an insulator a semiconductor.

1. The gap between the highest full band and the next empty band may be very small. In this case it will be sufficient to heat the material a little – sometimes room temperature will suffice – for the thermal energy to transfer some electrons across the gap into the empty band. In this context the empty band is called the conduction band, because once electrons are in it, they are free to receive further energy from an electric field and can, therefore, conduct electricity.

The more we raise the temperature, the more electrons get into the conduction band. Therefore, the conductivity increases with temperature, a fact discovered experimentally for silver sulphide a hundred years before it could be understood on the basis of a general theory. A semiconductor which conducts only by virtue of electrons thermally excited into the conduction band is called an intrinsic semiconductor. We may define a class of materials called semiconductors as materials with an empty conduction band but a small forbidden gap.

2. The second way by which electrons can be introduced into an otherwise empty conduction band is by the addition of small amounts of certain impurities. If the impurity atoms are of the right type, they will lose an electron fairly readily and this electron will enter the conduction band, leaving a positively charged localised centre behind. An impurity which behaves in this way is called a donor centre, because it donates electrons into the conduction band. In this way it is possible, to a limited extent, to make true insulators into semiconductors, but the process is more commonly used with real semiconductors.

3. The third and final way in which conductivity can be obtained in an insulator or enhanced in a semiconductor is by shining light of a suitable wavelength onto it. If conditions are right, then the energy of the light can be absorbed by an electron in the full energy band and this additional energy may transfer the electron across the forbidden gap into the conduction band. We may imagine the electron to be shot across the gap when it is hit by a photon. The process is known as photoconductivity and was extensively investigated by Pohl and his school. It is, of course, also extensively used in practical devices, such as light meters and photocells.

The neat enumeration given here may mislead the reader into thinking that once this theory was understood, in the mid thirties, all was clear and simple about semiconductors. Far from it. The very fact that impurities can alter the conductivity and the further fact that the lifetime of electrons in the conduction band, their velocity in an electric field, and many other parameters are critically dependent on purity and perfection of materials, made the unravelling of semiconductor problems extremely difficult and lengthy.

One further complexity ought to be mentioned. By measuring the effect of a magnetic field on a conductor or semiconductor, it is possible to determine whether the current is carried by electrons, with their negative charge, or by some positive charge carriers. To the bewilderment of early investigators, it was found that, in many cases, there is a contribution to the current by positive charge carriers. A further development of the band theory managed to solve this puzzle. When an electron is removed from the full band, as it is by thermal agitation in an intrinsic semiconductor, it leaves behind an unoccupied energy state. Once such an opening has been created near the top of the full band, other electrons in the band can gain a little energy by occupying the empty position instead of their own. In this way some possibility of absorbing energy from an electric field has been created, but the complex movement of the many electrons is such that it can be more readily described as the movement of a single positive charge carrier. Because the effect is caused by a hole in the otherwise full band, we call such a carrier a hole. A hole is a positive charge carrier caused by the collective movement of many electrons in a nearly full band. In an intrinsic semiconductor we have as many electrons as holes because each electron raised to the conduction band leaves behind a hole in the full band (often called the valence band). There are as many positive as negative carriers of charge. In a semiconductor containing donors, we may have as many electrons as there are donors, without any contribution from positive carriers. In this way we can obtain a semiconductor in which the majority of carriers is negative and which is called an n-type semiconductor. It is possible, however, to introduce a different kind of impurity, one that is essentially lacking an electron. When this is the case, the impurity will readily attract an electron from the valence band and will hold it in a localised position. The electrons thus accepted into these localised impurities, called acceptor states, will leave holes behind them. A semiconductor in which the majority of charge carriers is created by acceptor states will be a p-type semiconductor, because the majority-carriers are positive.[31, 32]

The unravelling of all these things took a very long time and has even now been accomplished in complete detail for only a limited number of materials. One obstacle to understanding was the difficulty of producing pure materials. The requirements of purity for semiconductor work were far beyond the scope of chemistry in the thirties. Another obstacle was the difficulty of growing crystals of adequate quality. We have already seen that it took several years for Pohl's research group to obtain good crystals of rock-salt and similar simple substances. To grow crystals of high melting point materials, such as silicon, proved infinitely more difficult and good silicon crystals did not become available until the fifties.

While all this theoretical work was going on and while a few experimenters continued to clarify the concepts and measure the parameters relevant to semiconductors, technologists experimented with a number of devices without bothering too much about their theoretical intricacies. In some cases they produced successful devices: in others, devices had to wait for a better understanding of the underlying physical processes.

Amongst the devices in common use in the thirties and forties were a variety of photocells and small current rectifiers, made from either copper oxide or selenium. Such rectifiers were very different from the cat's whisker diodes used for early radio detection. They consisted of an array of copper discs with one side oxidised and the other not. This gave a series of copper to copper oxide contacts which would rectify quite effectively, but whose resistance was too high to carry a very large current. The rectifiers were adequate for all electronic circuits and were used whenever a vacuum diode was inconvenient, because of either its size or its power consumption. The copper oxide could be replaced by selenium, so that the rectifier consisted of a series of selenium to metal contacts.[33]

Copper oxide and selenium rectifiers were manufactured in large quantities despite the fact that the theory of their operation was far from clear. The existence of these devices presented a sufficient challenge to physicists to produce a number of publications on the theory of rectification in the late thirties. The main contributors were Mott, Schottky and Davydov and the theory explained, essentially correctly, the rectification of an alternating current flowing through a metal semiconductor contact, such as selenium–metal. Essentially, the theory postulates that the semiconductor is depleted of current carriers near the contact with the metal. This forms a barrier to prevent the flow of electrons in equilibrium. When a voltage is applied in a direction so as to increase the width of the barrier, no current will flow. When the applied voltage decreases the barrier, current can flow across it. In this way a rectifying contact is obtained. The theory of metal semiconductor contacts was more or less complete by the beginning of the Second World War.[34]

Selenium and other semiconductors were also in use as photocells; devices used either to measure light intensity or to actuate some electrical equipment when illuminated. Light meters for laboratory and photographic purposes were available in the inter-war years and automatic doors were among many other light ac-tuated mechanisms in use, on a limited scale, in the thirties. All these photoelectric devices use one of two phenomena; either the change in conductivity of a semiconductor when light falls on it, or the emission of electrons from a surface under the same circumstances. The latter can be obtained from a metal, but the former is a special characteristic of semiconductors and insulators. Only a substance which is short of electrons will alter its conductivity substantially when free electrons are made available by shining light of a suitable wavelength onto the substance.

Throughout the thirties and forties, the dream of achieving solid state amplifiers persisted. Numerous inventors and investigators, using a variety of materials, tried to achieve amplification by one of two ways; either by what became known as the field effect, in which additional charge carriers are induced in the semiconductor by a voltage applied to an adjacent electrode, or by some modulation of a current injected into the semiconductor by a suitable contact.

The reasons for the continuing dream are not hard to find. After all, electronics had become a major industrial activity and any im-provement on the vacuum valve could revolutionise electronics and give very large economic returns. That improvement was desirable is obvious from the many real weaknesses of valves. They are large, fragile, have a relatively short working life and, worst of all, con-sume a lot of power. This last fact adds considerably to the bulk of any instrument using valves, because it must incorporate a transformer and associated rectifiers and smoothing circuits to provide the heating current for the cathodes and the d.c. supply for the anode. Such requirements were particularly troublesome for portable devices, such as radios and walkie-talkies.

Apart from the purely scientific demonstration of amplification in a crystal by Hilsch and Pohl in 1938, the best known invention of a crystal amplifier is that by Lilienfeld, as early as 1925. Lilienfeld applied for a number of patents for a complex device vaguely akin to a transistor. It was a multi-layer structure consisting of metallic and semiconductor layers.[35] The argument whether or not it could have worked is an interesting one and crops up in the scientific literature from time to time.[36,37] Lilienfeld led a somewhat obscure life and certainly did not have the means and facilities to test and develop his ideas.[38] Perhaps the course of history would have been different had he worked in a large industrial laboratory, but most likely his ideas were not practicable simply because of lack of knowledge of solid state physics in his time. In some fields it seems possible to invent practical devices without much theoretical background: in the field of solid state amplifiers this has proved quite impossible.

A number of inventors attempted to produce an amplifier based on the field effect. There is a certain irony in this work. On the face of it, there could be nothing simpler and the first patent for such a device was filed in 1935 by Heil.[39,40] The same idea was tried in the Bell Laboratories in the immediate post-war period, advocated mainly by Shockley, and failed. This early failure eventually led to a different device, the junction transistor, and also to further insight into the problem of semiconductor surfaces. Ultimately, years after the initial announcement of the transistor, the field effect transistor became a reality. Indeed it is now in common use and is a key element in modern integrated circuits. This is a case of a very large gulf of ignorance separating the early ideas from any possibility of realisation. Much knowledge had to be accumulated before the original ideas could be modified so as to create practical devices.

The idea of the field effect solid state amplifier is simplicity itself. Take a slab of semiconductor material. Remember that it conducts electricity poorly because the number of its carriers of electricity, electrons or holes, is severely limited. Make the slab one plate of a condenser and the other plate simply a metallic electrode. Apply a voltage to the metallic electrode and this will induce additional charge carriers into the semiconductor. In this way the conductivity of the semiconductor can be modulated. If the circuit is correctly arranged, a low power signal on the metallic electrode can cause a much higher power change in a circuit containing the semiconductor as a current limiting element. The idea is beautifully simple, but for many years it refused to work. Not until 1947 did Bardeen discover why. The additional charge carriers do not contribute to the conductivity of the semiconductor because they become immobilised by being trapped at the surface of the crystal.

In the patent by Heil, which may be regarded as a direct precursor of Shockley's idea for a semiconductor amplifier, the materials suggested were tellurium, iodine, copper oxide and vanadium pentoxide. A somewhat bizarre list in the light of modern knowledge,

but at least some of these materials might, with further development, have worked. At the time, further effort was not made.

Several other inventions were made during the thirties, but none
of them led to any technical breakthrough. With the beginning of
the war, all semiconductor effort was increased and directed towards
two aims; producing better detectors for microwaves and understanding better the properties of the main semiconducting
materials in order to increase the possibility of creating new devices.

CHAPTER 3

Science after the Second World War

Science emerged from the war as an entirely new entity. No longer was it a pastime for gentlemen or an obscure occupation for the bright but useless. The war had caused science to emerge from obscurity into the limelight. Suddenly, science was big, powerful, almost omniscient. Scientists became oracles and saviours.

The Western World emerged from the war with a strong faith in a better, more prosperous tomorrow, and the tool to achieve it, the fulcrum of the effort, was to be science and science-based technology. Science emerged from the war as a new faith, but it was a faith founded on sound experience. The First World War has been termed the chemist's war, but chemists somehow did not capture the popular imagination and did not create the image of a war won by scientists. The gruesome picture of men slaughtered in the trenches by mostly quite primitive means was the predominant impression left by the war. The Second World War was different. Quite early on the marvels of science began to emerge. Complex aeroplanes, supported by highly scientific radar, appeared to save Britain in the early days of the war. Even more complex aeroplanes, supported by even more sophisticated radar, continued to play a major role.

But it was not only scientific hardware which played an obviously decisive role in the war. Scientists became the backroom boys, the boffins, of the military leaders, and this was the most gruesomely practical role they had played. Scientists, very well-known highly abstract thinkers amongst them, became advisers on strategic matters and in the process invented the new science of operational research; the method of applying scientific thinking to the solution of practical, initially operational, problems.[1]

As if radar and operational research were not enough, scientists pulled off the greatest and most horrifying of all publicity stunts by exploding two nuclear bombs at the end of the war in the Far East. With these spectacular fireballs and the no less spectacular loss of life caused by them, the scientists' power achieved Promethean dimensions. If science and scientists had the power to make such decisive and spectacular contributions to the deadly serious business of war, surely their powers could be harnessed to the more joyful business of building a new and better peaceful world.

Although the number of scientists graduating from universities had been slowly increasing during the thirties, their number grew explosively after the war. It became part of the higher expectations of the post-war world to receive a university education, and a very high proportion of the post-war generation of students wished to study science. Science had become powerful, had become a useful and gainful profession, and, above all, science was exciting. There seemed no limit to the possible achievements.

It was in this atmosphere of hope and expectation that Bell Telephone Laboratories assembled their team of scientists to work on solid state physics problems, with the hope of achieving devices useful to the telecommunications industry. Their work was to be based on the war-time work in both the U.S. and Great Britain on devices used for radar detection.

During the war, when reliance on unreliable silicon detectors was a necessity, great efforts were made to improve knowledge of those devices. It was at long last realised that an important factor, if not the key factor, was an understanding of the materials used. Better devices could be made only if the materials used could be better controlled and their quality improved and consistent.[2]

The two materials picked for further investigation during the war were germanium and silicon. The reasons for this choice are not entirely clear, but it would appear that reasonably good performance of empirically obtained radar detectors in these materials was the main factor. By preparing a polycrystalline specimen of either of these materials and applying a tungsten wire to the surface, it was possible, after careful probing for a sensitive spot, to obtain a reasonable detector. The performance could be improved by tapping and by 'forming', achieved by discharging a small condenser into the contact.[3] There were other reasons for the choice of silicon and germanium. These are elemental semiconductors and therefore the chemistry of preparation was expected to be relatively easy. Only the one element needs to be retained and every other element removed, as far as possible, as an impurity. This is not the case with compound semiconductors, where one of the problems of preparation is to keep the proportions of the constituent elements correct. A further important consideration in the choice of these materials was their robustness. Both silicon and germanium are hard, strong, if somewhat brittle, substances of metallic appearance. There is no problem of crumbling when they are handled or evaporating under tropical conditions.

Having decided to investigate germanium and silicon as promising materials for use in radar detectors, the Americans set out to investigate them on a large scale and in a systematic manner. While the manufacture of radar detectors went on, especially in Great Britain, and while radar was in operational use in every theatre of war, the Radiation Laboratory at the Massachusetts Institute of Technology (MIT) was given the task, at the U.S. Government's instigation and expense, of organising research into the fundamental properties of germanium and silicon.[4]

Much of the work was distributed, in what appears to have been a fairly haphazard manner, amongst a number of research laboratories outside MIT. Many universities and industrial organisations were asked to work on these materials. It seems that the British and perhaps even the Germans had made more headway in research into semiconductor rectifiers by 1939 and that American work was, at least initially, based on British investigations into such phenomena as the hot spots on polycrystalline silicon.

> In the radar programme, we did rely initially on the British early development in that field. You had the lead on us at that stage.
> V. A. Johnson, Purdue University

The scale and dedication of the American semiconductor research effort was such that this situation was unlikely to continue. As many as thirty or forty U.S. laboratories were delegated the task of examining semiconductors for use in radar. The work at just one of these – the Physics Department at Purdue University – may be taken as an example.

The work at Purdue began early in 1942 under the leadership of Karl Lark-Horovitz, who had expressed a personal interest in that material, though his previous interests had been in X-rays and structures and in nuclear physics. Purdue approached the semiconductor problem afresh, having been entirely concerned with the spectacular developments in nuclear physics of the thirties. They decided to start by investigating purification procedures for germanium. All the chemical techniques known at that time were used, such as hydrogen reduction and melting in an inert gas atmosphere. This early work led to the production of reasonably pure, reasonably large polycrystalline ingots, from which detectors could be fabricated. The first detectors were made in the summer of 1942. At the same time, Lark-Horovitz declared what Wilson had long felt, that germanium and silicon were not simply impure metals, but intrinsic semiconductors even in their purest state. Nobody had clearly stated this before; the question had always somehow been left unanswered, or even unasked. The team at Purdue consisted of about a dozen scientists dedicated to increasing knowledge of germanium so that it might be used more effectively as a radar detector, but in the course of their investigations, at least two of these men came close to discovering the transistor.

One of the problems associated with practical detectors was burnout; the device became useless after an overload of current. New techniques of fabrication were investigated to produce detectors more resistant to burn-out. During this investigation the rectifying properties of the diodes were studied most carefully and gradually it became possible to achieve higher forward current, smaller reverse current, and the ability to withstand larger voltages.

A strong coupling between theory and practice was evident even in those early days. At Purdue the work was divided into three groups. One group investigated the fundamental electrical properties of germanium, such as mobility of charge carriers, Hall effect, and resistivity as a function of temperature and of impurity content. The second group was employed on preparing purer and better materials and the third group was working on actual detector devices. By October 1943, this last work had paid off in the sense that it was now possible to fabricate detectors which could withstand 150 volts in the non-conducting direction. This was regarded as a considerable achievement. The war-time achievements of the Purdue group are summarised by Lark-Horovitz:

> At the end of the war the Purdue group had (a) shown electrical properties to be predictable from impurity content, (b) predicted resistivity and thermoelectric power in the range of temperature available at this time (down to liquid air temperature) from the number of electrons given by Hall effect measurement, (c) determined the mobility ratio for holes and electrons. First infrared measurements by K. Lark-Horovitz and K. W. Meissner yielded the

dielectric constant for Si 13 for Ge 16–17. Highback voltage rectifiers were perfected and the present type cartridge introduced by R. N. Smith. Methods of melting and production of high-purity ingots brought to high perfection by R. M. Whaley. The group decides to abandon development of detectors and the practical applications and to concentrate primarily on the basic investigation of germanium semiconductors.[5]

This may not seem much to show for nearly four years of dedicated effort by a group of something like a dozen graduate scientists. Yet the scientific community was satisfied that the work constituted a very great achievement and was carried out with devotion and competence. Science is a painstaking business.

Despite the great success of the coupling of practical and 'fundamental' work during the war, the Purdue group reverted to type and decided to go back to fundamentals as soon as the national emergency was over. Had they not done this, had they remained alert to practical possibilities, it may be speculated that they, rather than Bell, might have carried off the great prize of the transistor. But they were scientists, not inventors, and they did not have any coupling with potential markets. They did not want sales: they wanted scientific prestige.

Even after all these years, there was still talk about ingots, meaning polycrystalline materials. The art of growing single crystals, so essential for later developments, was developed only after the war. Even so, war-time concern with purifying and understanding semiconductor materials was crucial to early post-war success in obtaining the transistor.

> If it hadn't been for microwaves or UHF radar, we probably wouldn't have needed crystal detectors. If we hadn't gotten crystal detectors, we probably wouldn't have had a transistor, except that it might have been developed from some entirely different pathway.
>
> Quentin Kaiser

> I think the thing that really made the subject take off was something that happened during the war, namely the use of germanium and silicon ... for microwave detection ... which resulted in progress of the materials technology of silicon and germanium and which then provided a basis for the development of the subject on a combined basis of theory and experiment after the war. The big problem was that while there was substantial development of theory during the 1930s, there was little substantial contact, I would say, between theory and experiment.
>
> Harvey Brooks

Without these advances in materials, you wouldn't have stood a ghost of a chance of making a transistor.   J. R. Pierce

Indeed, it may be argued that the success of Purdue with germanium, which was greater than contemporary success with silicon, determined the early predominance of germanium as a semiconductor material. The more difficult silicon problem was resolved in later years and germanium eventually lost its early lead to the superior qualities of its rival. But who knows what materials we might be using today if these two had not established such a substantial lead during the war, a lead that could have been challenged only at enormous cost and which was never seriously rivalled by the dozens of compound semiconductors proposed at various times. These now have minor specialised applications, such as the gallium arsenide light emitting diode, but in electronic amplifiers and circuitry silicon reigns supreme.

With science firmly established as a real force in the world, and with germanium and silicon firmly established as useful device materials, the era of large-scale developments in solid state electronics had begun.

CHAPTER 4

The Bell Laboratories

The 'major invention of the century'[1] occurred at the Bell Laboratories in Murray Hill, New Jersey on 23 December 1947. The transistor, a device which used semiconductor material to amplify an electrical signal, was discovered on that day, its most immediate impact being to ensure a merry, if busy, Christmas for those most concerned with the invention.[2] So important is that invention that it has become certainly traditional and almost mandatory to examine its source, Bell Laboratories. The justification for this is the argument that as Bell have produced the transistor and many other significant inventions, then Bell methods of invention are worthy of consideration for use elsewhere.[3] There is a certain logic in the reasoning that methods which have produced much new knowledge are likely to be the best to produce more new knowledge; though there is also something paradoxical in the thought that there should be established methods of creating the revolutionary. A graceful sidestep away from generalisation will avoid collision with the theorists of invention and innovation, and we shall only tackle the specific question why it was that the transistor was invented at Bell.

The Bell Laboratories are massive. They employed 5700 people in the late forties, of whom over 2000 were professional staff of the highest quality.[4] They now have about 17 000 staff producing, *inter alia*, some 2300 scientific papers and some 700 patents annually.[5] It is easily the largest industrial research organisation in the world. Bell is the research wing of the American Telephone and Telegraph group of which Western Electric is the opposite, manufacturing wing. The group is concerned with communications and Bell Laboratories with ways in which communications systems can be improved. The scope of such work, ranging from telephones to satellites, is rendered yet wider by the interest Bell shows in research into all sorts of matters peripheral to communications. The problem of recycling plastic from old cables is an example.[6] Moreover, Bell is responsible for research into the long-term needs of telecommunications to cope with situations which might not arise for another twenty years, if then. Not surprisingly, such work often involves the Laboratories in matters which are remote from even the wide field of telecommunications. To pay for this massive programme, Bell now spends about $500 million a year, of which roughly $300 million comes from Western Electric and the remainder from American Telephone and Telegraph.[7]

As an industrial research organisation, the record of the Bell Laboratories is superb. Whatever index is used – numbers of scientific publications, of patents, of major inventions, of honours and awards – the scientific contribution of the Laboratories has been outstanding.[8] Bell physicists, for example, published many more papers in the prestigious *Physical Review* than those from any other industrial organisation in 1956–7. In fact, Bell provided a quarter of the total industrial contribution and ranked sixth among all contributors.[9]

Despite this record, the near monopoly of the Bell System over American telecommunications renders it liable to censure under anti-trust legislation. In 1956, such criticism resulted in a Consent Decree by which Bell was forced to surrender its existing transistor patent rights. Bell is again under attack from the same quarter and is involved in litigation which may last into the eighties.[10] Perhaps it is because Bell feels vulnerable to such criticism that it has always stressed the value not just of research but of the very basic research it does. Though the first American frontier has disappeared, to paraphrase a fairly typical piece of Bell publicity, there remains the new and exciting frontier of science waiting to be tamed by basic research.[11] Much effort has been dedicated to propounding the Bell art of converting science into technology[12] and very often the invention and development of the transistor have been used as evidence of how effective the technique employed at Bell Laboratories can be.[13]

Fundamental to the question of why Bell, rather than any other organisation, discovered the transistor is an appreciation of the sheer size of Bell Laboratories and their reputation as a research organisation equal in standing to the best universities. This enabled the Laboratories to recruit the cream of available talent in all disciplines.

The organisation of the Bell Laboratories allowed for frequent and informal contact among scientists of different disciplines, a factor which is important when invention is dependent upon simultaneous progress in several fields, and when progress in any one field may be triggered off by developments in another. The invention of the transistor was a product of the collaboration of physicists, chemists, metallurgists and engineers; it required just the sort of inter-disciplinary co-operation that Bell encouraged.[14] It appears that the art of management of research and development consists of striking the right balance between giving scientists sufficient scope to develop their ideas and their enthusiasm and directing the work towards a goal acceptable to the organisation. The Bell management were true masters of this art.

Yet it would be wrong to mistake the high academic standard of Bell and the university atmosphere of the Laboratories for the sort of academic freedom normally found in universities. Bell has sometimes claimed not only that this freedom exists,[15] but that it was particularly relevant to the invention of the transistor.[16] At least one scientist working at Bell at this time talks of a very strict hierarchy with freedom only for those at the top. 'The humble Ph.Ds . . . felt like cogs in a very large machine' and there was difficulty convincing those who made decisions that a particular line of research was worth following.[17] There are grounds for being severely sceptical of any claim that the invention of the transistor was the product of scientists allowed the freedom of loose reins. Industrial research organisations do not work that way – not even Bell Laboratories.[18] In any case, scientific freedom in research can be a sterile luxury without funds to support that research. It has been estimated that Bell Laboratories spent $140 000 on inventing and developing the transistor between 1946 and 1950,[19] though another, probably more realistic, estimate puts the figure at just under $1 million for the three years up to 1948 alone.[20] Scientific research is expensive and involves investment. In an industrial organisation, such investment involves a degree of commitment to the project and, almost inevitably, a degree of hierarchical control over that project. The huge size of the Bell Laboratories, its very catholic interests, its efforts to decentralise management and its exceptionally long-term research efforts no doubt might reduce demand for cost-effectiveness, but some demand for value must always have existed and certainly exists today.[21]

One of the reasons for granting the top scientists a large degree of freedom in formulating and pursuing research proposals is the idea of statistical chance of success. A single long shot may or may not pay off. With many such shots, the chance of any single pay-off is much higher.

> We found it was simply productive to have people engaged in pure research which did not necessarily have any apparent application in the Bell System, for if you had enough of this there was sufficient fall-out to pay for the whole effort many times over again . . . You can never say which developments are going to be profitable.
>
> Dieter Alsberg

The three men who played the leading role in the invention of the transistor and who were later awarded the Nobel Prize for their efforts were John Bardeen, Walter Brattain and William Shockley. All three were physicists, though their approach to the subject differed. Bardeen was always the pure theorist and Brattain, at the other end of the scale, the experimentalist, with Shockley, another theorist, perhaps somewhere in between. Shockley's habit of inventing new terms and using esoteric analogies posed some difficulties, for both Bardeen and Brattain often found themselves having to interpret his language.[22]

Walter Brattain was the first of the three men to work for Bell Laboratories. When he joined in 1929, he was assigned to a branch of valve research dealing with surface properties of metals. By 1931 he had been allotted a new task, studying the phenomenon of rectification in copper oxide.[23] Copper oxide is a semiconductor, albeit a very complex one and exceedingly difficult to work with. When William Shockley joined Bell in 1936, he too was first put to work on valves, though it seems he would have chosen to work closer to his own field, the behaviour of electrons in solids.[24] Certainly one of Bell's main research interests at this time was the valve, but it may be more than coincidence that both young Ph.Ds. were made to serve their time in the valve department and were not at first allowed to follow their own inclinations into fields in which they would have felt more at home.[25]

The Director of Research of Bell Laboratories at this time was Mervin Kelly, a man whose long-term goals apparently made a great impression on Shockley.[26] As early as 1936, Kelly felt that one day the mechanical relays in telephone exchanges would have to be replaced by electronic connections because of the growing complexity of the telephone system and because much greater demands would be made on it.[27] As this is hardly technically feasible using valves, it seems that Kelly was thinking not simply of a radically new valve technology, but perhaps of radically new electronics.[28] This was how Shockley interpreted Kelly's ideal, though whether either he or Kelly felt that this new electronic device was to be a switch or an amplifier is not clear. It seems most likely that Kelly saw the logical progression from a semiconductor rectifier in copper oxide to be a semiconductor switch rather than an amplifier.

> Shockley and his group were trying to measure what was going on in a rectifier, so that they could develop a better rectifier for telephone switching. At that time the latest state of the art in telephone switching was by simple on–off devices – reed relays and the so-called crossbar system – and what they were looking for was a solid state on–off device, a simple switch.   Maurice Apstein

> I don't think that when Kelly made his pronouncements [about electronic switches] in the 1930s, that anybody at Bell Laboratories foresaw the direction in which the thing would go. They just felt it was important.   Harvey Brooks

> The driving force for the transistor was ultimately a need . . . there was a clear-cut need to get something that was an improvement over the vacuum tube . . . It wasn't basic research just for the sake of basic research. It was basic research to provide a major problem solution for something in the communications mould.   Herbert Kleiman

It may be that others at Bell Laboratories interpreted Kelly's aims rather more ambitiously and that even their limited experience with vacuum tubes had suggested that radical change was unlikely to emerge from that quarter. In the late thirties, both Shockley and Brattain were turning more and more to their own fields of expertise, to the behaviour of electrons in solids, in the hope that a revolutionary approach to the problem would yield the revolutionary solution it demanded.

In 1939, Shockley and Brattain collaborated in an attempt to produce a semiconductor amplifier using copper oxide, a choice of material stimulated by experience with copper oxide rectifiers.[29] The idea was to insert a tiny controlling grid into the oxide layer on the copper in the hope that it would control the current passing through the semiconductor, much as the grid in a valve worked.[30] Slightly later experiments used various copper plates at the surface of the copper oxide in attempts to produce an amplifier based on charges induced in the semiconductor. This type of device is now known as a field effect transistor. All these attempts were total failures. The failure of the grid hardly surprised Brattain at any rate, for he and J. A. Becker had convinced themselves of the impossibility of making a transistor by inserting a grid into the semiconductor a good year before Shockley 'amused' them with his determination to experiment along these lines. Shockley thinks that Brattain slightly misinterpreted his intentions. The result, however, was the same – the device did not work.[31]

> Anybody in the art was aware of the analogy between a copper oxide rectifier and a diode vacuum tube and many people had the idea of how do we put in a grid, a third electrode, to make an amplifier.
>
> Walter Brattain

This was not the only work on semiconductors going on at Bell during the immediate pre-war period. A team including R. S. Ohl, J. H. Scaff and H. C. Theuerer – chemists and metallurgists rather than physicists – was working on ways to obtain purer silicon for use in detecting the shorter radio waves. Their work resulted in both a greater ability to control the purity of the silicon and in a crude capacity to create n-type or p-type silicon,[32] accomplishments that heralded the vast progress that was to be made in this direction during the war.

One slice of silicon with a particularly sharp boundary between the n-type silicon at one end and the p-type at the other was shown by Ohl to Brattain and others in 1940. They were astounded that a torch shone on the junction should produce ten times the expected electromotive force and Brattain at first suspected trickery.[33] It seems that Bell's interest in silicon detectors of short waves had first taken practical form some years before this experiment, with the visit of G. C. Southworth to a radio market in New York where he had ferreted out some obsolete silicon detectors of the cat's whisker variety.[34] In view of the vital importance of wartime work in making the essential preparations for the invention of the transistor, it is interesting how much work of this sort had already taken place at Bell during the thirties, albeit at a somewhat casual and desultory pace. With hindsight, it is clear that this work lacked the direction which would have been given by a proper theoretical framework.

During American involvement in the war, the main semiconductor research at Bell was on silicon, but was very much subservient to the national research into semiconductors for radar detectors organised by the Radiation Laboratory at MIT. Brattain and Shockley did not participate as they found themselves assigned to submarine work outside the Bell Laboratories. They were both concerned with the use of radar for submarine detection, but worked separately and were forced to postpone their interests in semiconductor research.[35]

While some work on silicon continued at Bell during the war, the developments with germanium at Purdue are, at this point, of more interest. Seymour Benzer and Ralph Bray were both graduate students in 1944, studying under Lark-Horovitz for their master's degrees. Benzer was interested in the high resistance of point contacts on germanium in the reverse direction: Bray in the forward direction. Investigation involved putting a point contact wire onto a piece of germanium and measuring the resistance of this rectifying contact (spreading resistance). The results defied existing theory by being much too low. What neither knew was that the injection of minority carriers was responsible, but there was then no recognised concept of minority carrier injection, little literature on semiconductors and not even courses on solid state physics.[36]

While it has been suggested that the end of the war brought little change in the direction and intensity of semiconductor work,[37] Professor Bray feels that there was a distinct resurgence of interest in nuclear physics at Purdue. One man who might have shed some light on the findings of Bray and Benzer was shifted to build a linear accelerator. 'And so there we were left, just two graduate students on a project with absolutely no assistance from anyone else.' In fact some real concern was shown in their work by an interesting source. Bell had maintained contact throughout the war with the work at Purdue through the MIT co-ordination, but immediately the Japanese surrendered, William Shockley and Stanley Morgan visited Purdue to find out what they could.[38] Shockley's interest was, even at this early date, the creation of a solid state equivalent of a triode. In this, his concern was rather different from that of Bray. 'I was interested in explaining the effect [the low spreading resistance] that I was seeing, but I wasn't particularly interested in the idea of a solid state triode.' By 1947, Benzer had left for Bio-Physics and, according to Bray, 'That just left me and occasionally visitors would come and I'd tell them about this and everyone would scratch their head, but there was no explanation for these phenomena.'

Bray and Benzer reported their findings on point contacts on germanium to a meeting of the American Physical Society early in 1948, only weeks after the invention of the transistor at Bell, but still months before the secret was announced. Walter Brattain was in the audience, knowing very well that the phenomenon was due to minority carriers and realising how close Bray and Benzer had come to discovering the transistor. As Bray said later, 'What's perfectly clear is that if I had put my electrode close to Benzer's electrode . . . we would have gotten transistor action.' At the time, Bray sought out Brattain who was then very worried that Bell would be beaten in the publication of a scientific paper on the transistor. Relating his conversation with Bray, Brattain recalls, 'I just let him talk, saying nothing. Bray finally said, "You know, I think if we would put down another point on the germanium surface and measure the potential around this point, that we might find out what was going on." And I couldn't resist saying, "Yes Bray, I think that would be probably a good experiment!" and walked away.'[39] Ralph Bray had just described the very experiment which had led to the invention of the transistor at Bell only a few weeks before.

> The amazing thing which I could never understand is why Purdue didn't invent the transistor.   Henry Levinstein

In 1945, Brattain and Shockley were joined at Bell by John Bardeen. Bardeen had interrupted academic research on the theory of metals with five years at the Naval Ordnance Laboratory in Washington and was anxious to return to solid state physics. It was Shockley who was partly responsible for persuading him to work at Bell rather than a university.[40] Bell was reorganising after its wartime research and a new group was set up to conduct a fundamental investigation into those aspects of solid state physics most relevant to communications equipment. The authorisation for this work was signed in the summer of 1945 by Mervin Kelly, who at that time was vice-president for research at Bell, and resulted in the organisation of a solid state physics sub-department.[41] In charge of this group were William Shockley and Stanley Morgan, the two men who had been keen to see what was going on at Purdue. It seems that even at this stage there was a general realisation in the group that possibly the most important development that could arise from Bell's solid state research would be a semiconductor amplifier.[42]

In fact, the optimism that such a thing was possible rested on very shaky foundations. Kelly's ambition to replace mechanical relays in telephone exchanges with electronic ones may have been an important driving force, but it was still no more than an idea.[43] Although he did not follow the work in detail, Kelly inspired at least some of those working under him.[44] Ohl had produced amplification in a semiconductor towards the end of the war, but this was a result of resistance varying with temperature – the thermistor rather than the transistor effect – and produced a totally unreliable device.[45] Shockley still saw the field effect transistor as the likely answer, but attempts to produce such a device after the war met with no more success than those before the war.[46] In theory the field effect transistor should have worked: in practice it was a dismal failure and Brattain and Bardeen were set the task of discovering why.

The semiconductor group at Bell had deliberately chosen silicon and germanium with which to work as these were the simplest semiconductors and those to which so much attention had been given during the war.[47] The assumption was that when these materials were used in a field effect arrangement, the induced charge carriers would be free to move and thus contribute to the conductivity of the semiconductor. An explanation of the failure of Shockley's field effect experiments, suggested by Bardeen, questioned this view and proposed a theory of surface states which would immobilise the induced carriers. Bardeen postulated that electrons were being trapped at the surface of the semiconductor and hence no application of an electric field would substantially alter the number of free charge carriers within the whole body of the semiconductor.[48] If this were happening, it was important to find out more about it and consequently the semiconductor group abandoned attempts to make an amplifying device in order to concentrate on research into Bardeen's surface states.[49] It was during the exploration of the semiconductor surface with two closely spaced electrode wires that Bardeen and Brattain discovered that a small positive charge on one electrode would inject holes into the semiconductor surface which would greatly increase its capacity to carry current. This result gave Bardeen the idea that an amplifier could be achieved by closely spacing two wire electrodes on a germanium crystal. Brattain tested the idea and soon created the first point contact transistor. It was demonstrated for the first time on 23 December 1947 (Fig. 4.1).[50]

There are one or two points of importance which emerge from the way the transistor was discovered. Firstly, one is struck by the

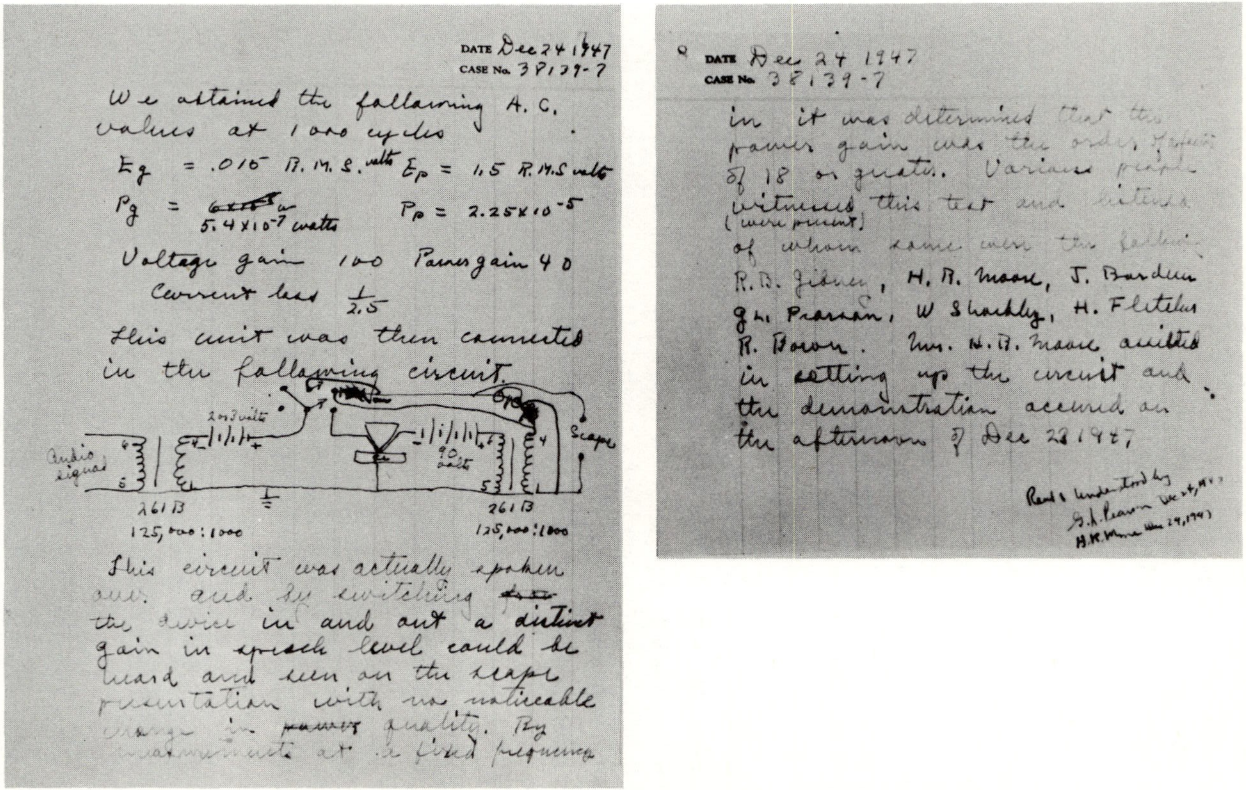

Fig. 4.1. Brattain's laboratory notebook recording the discovery of the transistor on 23 December 1947. (Courtesy of Bell Laboratories)

successful collaboration of Bardeen, the theorist, and Brattain the experimentalist. Brattain readily admits that he 'had an intuitive feel for what you could do in semiconductors, not a theoretical understanding'. The traditional tendency to value the latter and ignore the former may not be wholly justified. It is also interesting that, at least in specific detail, the transistor was the result of failure to create a field effect semiconductor amplifier and abandonment of the original project.[51] Though Shockley, the group leader, had been certain of the correctness of his theories for a field effect transistor, the problem of surface states had arisen, had totally changed the direction of research, and, amazingly, had resulted in a sort of amplifier he had never envisaged.

It had been Brattain and Bardeen who had invented the point contact transistor,[52] and neither Shockley nor Bell was particularly pleased that the transistor had come about in this way. When Bell came to patent the point contact transistor in 1948, the patent was granted to Bardeen and Brattain without any difficulty. Shockley, on the other hand, was unable to obtain a primary patent on his field effect principle because of work carried out before the war by such men as Pohl and Lilienfeld.[53] Although the point contact transistor was invented by Bardeen and Brattain, all photographs of the transistor's inventors included Shockley.[54] Pictures of the trio of inventors (Fig. 4.2) are so familiar that they have become implicitly associated with what has been described as Bell's 'mom-and-apple-pie image'.[55] No doubt Bell management did its best to encourage maximum cooperation amongst its scientists; such cooperation is certainly a major factor in the success of the laboratories. But the transistor was, in some degree, a product of strife, perhaps inevitable in a group of strong personalities. When Bardeen left the Laboratories in 1951, it was at least partly because he disagreed with the direction his work would have taken. Bardeen wished to move on to superconductivity, a field in which he was to be awarded a second Nobel Prize.[56]

> Many of the things that the Bell Laboratories are proudest of now were done in spite of management.   Walter Brattain

It is inevitable that even the most enlightened and far-sighted management will reject some ideas which are later proved to have been good ones. Despite the obvious importance of control over semiconductor materials, Gordon Teal, the man whose materials

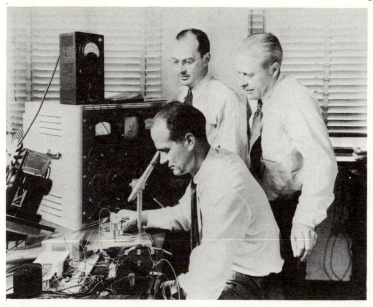

Fig. 4.2. The inventors of the transistor in 1948 – Shockley (seated), Bardeen (standing left) and Brattain. (Courtesy of Bell Laboratories)

research work was to lead to the pulling of single crystals from a melt of germanium and to the first commercial silicon transistor, received no support for the many germanium research proposals he submitted at Bell between February and August 1948.[57] Similarly, William Pfann, the Bell man who was eventually to develop the zone refining technique, absolutely critical in the construction of junction transistors, was actively dissuaded from pursuing his ideas.[58] Shockley, among others, did not think that zone refining would be important in the semiconductor field.[59] Again, the idea of James Early that the junction transistor of 1952 could be improved by making the collector thicker and the base thinner were accepted by Bell management only after a long, hard struggle.[60]

It is also unrealistic to see the transistor as the product of three men, or of one laboratory, or of Physics, or even of the forties. Rather its invention required the contributions of hundreds of scientists, working in many different places, in many different fields over many years. For example, one could examine the contribution of scores of chemists and metallurgists to an understanding of semiconductors and the development of their ability to produce them in a form useful to physicists.[61] Without this materials research effort, and particularly the acceleration it received during the war, there could have been no transistor.

William Shockley persisted in his belief that a better transistor than the point contact was possible. His understanding of semiconductor theory had all along suggested that this must be the case and the display of the significance of minority carrier action in the point contact transistor made him only more convinced. He decided to shelve his ideas for a field effect transistor and to pursue his theory for a junction transistor in which transistor action is achieved by sandwiching n-type semiconductor between p-type semiconductor. The theory had been developed by January 1948,[62] but the *Physical Review* refused to publish Shockley's report of the junction theory, its editor claiming that the quantum mechanics was not sufficiently rigorous.[63] Shockley's book containing the theory of this transistor was started in 1949, published in 1950[64] and the construction of the first reliable junction transistor was completed in 1951.[65] Amazingly, Shockley had formulated the precise theory of the junction transistor at least two years before the device was produced. The point contact transistor, with its delicately positioned electrodes, harked back to the triode valve and to the cat's whisker rectifier: the junction transistor, in which action took place within the body of the semiconductor, pointed the way to modern solid state electronics.

> Bill Shockley, who had a far better idea – the junction transistor – was at this time a bit frustrated because there was no simple way of making his invention. Walter Brattain had discovered the point contact transistor – John Bardeen's main contribution was in explaining the rather complex action going on – and nobody at that time could make the junction transistor Shockley had devised. I think that, at this stage, Bill Shockley may have felt that the true value of his contribution was not fully appreciated. In fact his early transistor patent was an intellectually very impressive achievement. Anyway, everybody started putting probes on pieces of semiconductor and studying processes like 'forming' which was a bit of black magic to make the transistors work better. I am sure Bill appreciated at that time that there was no long term future for point contact transistors but no realistic method of making junctions cheaply had yet been devised.
>
> Alan Gibson

CHAPTER 5

The transistor

The somewhat unexpected discovery of the transistor at Bell, just before Christmas 1947, posed something of a problem for the Laboratories. The discovery was obviously important and had to be patented, but it had been made during an investigation of surface properties and was not the result of an exhaustive programme to use the semiconductor surface to make a transistor. Consequently, there was a great deal to be done to find out precisely how the transistor worked and it was clearly desirable that as much of this as possible should be done before patents were filed. Until then, Bell had to keep the invention strictly secret. First patent applications were filed in February 1948,[1] though the transistor was to remain a Bell secret for about seven months in all. When D. A. Wright, then with G.E.C. in England, visited Bell in March 1948, he noticed a keen interest in British developments and remembers the Bell men trying to find out what he knew.[2] Even some of the Bell scientists asked to study the transistor were not told what it was they were working on.[3]

The first investigation undertaken in the endeavour to understand the transistor, was to use it as an oscillator to quell doubts that the device really could amplify.[4] It was also necessary to devise circuit applications and to find ways by which the transistor could be manufactured.[5] So unforeseen had the transistor been that neither Brattain nor Bardeen had thought of a name for the device. That honour fell to J. R. Pierce who deemed it appropriate that the name should join the series of existing semiconductor device names such as thermistor and varistor, 'and then at that time the transistor was supposed to be the dual of the vacuum tube, so if a vacuum tube had transconductance, this must have transresistance and so I was led to suggest transistor'.[6] The suggestion, originally made to Brattain, found immediate general acceptance.[7]

In view of the vast amount of work which had gone into semiconductors in recent years in so many laboratories, there was concern that Bell would be beaten in the presentation of a scientific paper.[8] Moreover, a single device was reasonably simple to make and it was not impossible that someone would stumble across the effect by chance. Work seems to have been at an advanced stage in France[9] and it has been said that a transistor was made in Britain within a week of the eventual Bell announcement in the United States and that 'it was an idea whose time really had come'.[10] There was also concern that Bell's plans to use the transistor might meet with some opposition from the Military, anxious to delay Russian scientific advance by restricting or classifying an important new invention. Though Bell was engaged in some military work after the war, the Military had not been involved in the invention of the transistor. Even so, Bell decided it was only prudent to announce the invention without first clearing it with the Military. To this end, the editor of the *Physical Review* was specifically asked not to reveal the contents of the scientific paper from Bell[11] before it was published in July 1948.[12] The Military had to make do with a preview of the transistor just a week before the public announcement in New York on 30 June.[13] It is generally thought that classification by the Military would have been difficult, unlikely and perhaps counter-productive as it would have grossly compromised Bell's position as a public utility working for public rather than military benefit.[14] From the mere knowledge that such a thing as a transistor was possible, there were perhaps twenty-five organisations which could have made one. Secrecy would have succeeded only in transferring the centre of what was to become a new industry away from the United States.[15]

On 30 June 1948 and during the months that followed, Bell publicity for the transistor was perhaps a little ahead of the technological development of the device.[16] Oblivious of the massive production and commercial problems that were to belabour the transistor, Bell proclaimed the transistor as a challenger to the vacuum tube, the simplicity of which would make for mass-production economies. An eight-foot model on wheels of the point contact transistor helped demonstrate how simple it was and the substitution of transistors for valves in a radio and a television set suggested a technical and commercial practicality that was not then justified.[17]

The announcement of the transistor excited so little reaction that it was a decided anti-climax to the months of secrecy at Bell.[18] Following the press announcement, the *New York Times* devoted all of 4½ column inches at the very foot of its radio chat section on page 46 to a most restrained account of the invention. The report was preceded by news that the programme 'The Better Half' now had a commercial sponsor and that station WNEW would soon broadcast traffic information.[19] Even the technical journals mostly managed to control their enthusiasm, many waiting until late 1948 to report the event, some until 1949 and some apparently not bothering at all. This seems strange considering the excitement of the transistor's inventors, but the explanation is probably twofold. At the time of the announcement, the transistor was little more than a laboratory curiosity. It was to be some years before the device was made to do something useful. In 1948, any prediction concerning the transistor's future could be little more than guesswork. But of more importance was the understandable, almost inevitable, predilection to view the transistor in terms of a familiar and apparently very similar device, the valve.[20]

Both the transistor and the valve were amplifiers and certainly at least some of the drive to invent the transistor had been prompted by a desire to find a better device than the valve. The very naming of the transistor had concentrated on its role as the dual of a valve and even the terms for the electrode wires in a point contact transistor – emitter and collector – were analogous to a picture of electron emission from cathode to anode in the common valve.[21] There seems little doubt that Bell thought of the transistor only as a replacement for the valve[22] and, consequently, it was hardly surprising that others saw the device in exactly the same light. Bell announced the transistor as the realisation of 'the old dream of a control valve for electrons flowing in solids',[23] and provided pictures of point contact transistors dwarfed by comparatively huge and clumsy valves.[24] Other pictures showed the transistor alongside paper clips or pencil points and inevitably stressed that attribute of the transistor, its small size, which seemed to be its greatest advantage over the valve.[25] There had been considerable interest in 'miniaturising' valves for years and it was quite reasonable to see the transistor as just a further, albeit large, step in the same direction. The technical press readily accepted this view and printed articles with such titles as 'Tiny device said to serve functions of vacuum tube'[26] and 'Semiconductor replaces vacuum tubes'.[27] Typical of these accounts were descriptions of the transistor as 'a possible substitute for the radio valve'[28] and as answering 'a question scientists have been pondering for many years – how to make semiconductors amplify and thus provide a simpler, more rugged, smaller device that can perform the functions of a vacuum tube'.[29]

> I think there was general confidence here that it would be a definite replacement for the triode [valve] ... The idea of it having more potential than that evolved over a much longer period of time.
>
> V. A. Johnson

In fact, it mattered little at this stage what was thought of the transistor because a vast amount of development work remained to be done if the transistor was ever to become a technological achievement rather than just a scientific curiosity. Work went on at Bell and at other electronics laboratories and it was not until October 1951 that manufacture of the point contact transistor started in earnest at Western Electric, the production division of American Telephone and Telegraph. By October of the following year, the transistor was in use in some oscillator units in the telephone system; by late 1952 some hearing aids contained transistors, and by March 1953 Bell was using transistors in an automatic telephone routing device called a 'card translator'.[30] By this time, of course, Shockley's junction transistor had been announced, though it was not then in production.

Bell had embarked upon a policy of public divulgation of its transistor knowledge. It may seem strange that having gone to great lengths to invent the transistor and then to keep it secret, Bell should want to share its technology. Yet, in April 1952 Bell held a symposium at which it was revealed how point contact transistors could best be made and what progress had been made towards the manufacture of junction transistors. Representatives of some thirty-five electronics firms attended, each organisation paying an entrance fee of $25 000.[31] This sum was later deducted from payments made to Bell by firms making transistors under Bell licence. Yet, even in 1952, transistors produced at Bell were few and often far from reliable.[32] What Bell hoped to do, was to improve the transistor so that it might become a really valuable device, particularly for the communications system, by encouraging other firms to tackle the problem. By publishing, holding seminars and licensing widely, Bell did its utmost to stimulate others to play a part in developing the device.[33] This accorded well with Bell's coveted image as a public utility, but it made good practical sense too. The more Bell appeared to be performing a public service, the less the danger from the anti-monopoly lobby. The more useful the transistor became, the more Bell would gain both from licence fees and from application of the transistor in telecommunications. Not least important, everything Bell revealed enhanced its prestige as a research organisation and this tended to attract more of the highest quality staff.[34]

Most of the companies at the first Bell commercial transistor symposium eventually produced their own transistors under licence from Bell. Other companies cared less for protocol and secured their transistor knowledge from RCA with whom Bell had a very comprehensive cross-licensing arrangement which included the transistor. There were still other companies which manufactured transistors for many years without benefit of anyone's training or licence.[35]

Throughout the early fifties both point contact and junction transistors were manufactured, but it became increasingly obvious that the junction transistor had far more potential for development and would, in the long run, be much more readily mass produced. It is doubtful whether the point contact transistor could ever have played a major technological role.

Bell had waived all licence fees for the first commercial transistor product. As a memorial to Alexander Graham Bell and to his interest in the deaf, Bell did not require royalties on transistors produced for hearing aids after 1954.[36] Soon after the announcement of the junction transistor, the large electronics firm, Raytheon, became the first supplier of transistors to hearing aid manufacturers. Only eighteen months later, over fifteen hearing aid firms were buying junction transistors from Raytheon.[37] But even at that time, the giant was being rivalled by a new forty-man operation called Germanium Products, based in a dilapidated building in Jersey City. Prophetically, its President declared, 'Trouble with the big companies is too many long-haired boys and not enough practical horse sense ... We expect to chase the vacuum tube price to hell and gone.'[38] Junction transistors were the first to be used in hearing aids in quantity, though sometimes in conjunction with vacuum tubes in an attempt to reduce the unwanted noise produced by many early transistors.[39] The advantages of the new hearing aids were that they were smaller, lighter and their batteries lasted longer. One of the earliest had an overall volume of about 45 cc and, with batteries, weighed about 90 grammes. In comparison, the contemporary British

National Health amplifier unit alone – without batteries and earpiece – had a volume of 130 cc and was of unspecified weight. Battery life was increased about eight times and annual battery replacement cost reduced from about $40 to $3. Initial price of the hearing aid was certainly greater – about $150 or $200 as opposed to $50 for a hearing aid using subminiature tubes – but lower battery costs were some compensation.[40]

Remarkable as these improvements were, they should be seen in the context of the very radical changes that had been taking place in hearing aids since the late thirties. Hearing aids were reduced in volume and weight very much more between 1938 and 1945, for example, than in the following eight years, despite the use of transistors.[41] The design of head-worn hearing aids was certainly encouraged, but the introduction of the transistor to hearing aids really did no more than permit the continuation of a trend towards miniaturisation that had been evident long before the transistor's invention.[42]

The transistor was used in hearing aids instead of valves because it had certain advantages over valves, though there were some disadvantages. It was, of course, very much smaller than even miniature valves; it consumed much less power; was more efficient than the valve and acted instantly because there was no filament to heat. Because there was no hot filament, there was none of the problem of dissipating the heat associated with valve arrays. The transistor was also, at least potentially, a mechanically rugged device. An exceptionally long working life of 70 000 or even 90 000 hours was forecast for the early transistor[43] and there were even some who imagined that because the transistor was solid state its characteristics were virtually permanent.[44] Unfortunately, the vulnerability of semiconductor surfaces to contamination meant that this was definitely not the case. Early transistors were not only noisier than valves, but they could handle less power, were more restricted in their frequency performance and were more liable to damage by power surges, radiation and high temperatures. It was difficult to design a transistor to give even an approximation of the characteristics required, and even harder to produce two transistors with the same characteristics.[45] Engineers concerned with these problems spoke of the 'wishing-in' effect, the result of those testing transistors having to hope for the right characteristics, and the 'friendly' effect where a wave to a transistor being tested would very likely be acknowledged on the oscilloscope.[46] In addition to these considerable disadvantages, the early transistor was also very expensive – in late 1953, the best transistors cost about $8 or about eight times the price of a valve[47] – and, at least during the first half of 1952, they were in rather short supply.[48]

> For that decade or so, from '53 to '63, we had no choice but to go with vacuum tubes because they did a better job, and up until that time they were cheaper. You could get a perfectly good vacuum tube for about 75 cents. U.S. electronics expert

Thus, the transistor was not generally regarded as an electronic miracle in the early fifties.[49] As a completely new sort of device it posed staggering problems, difficulties of a sort that had never been encountered with the valve. Whereas the valve had been improved considerably by the application of technology based on experience, the problems encountered with the transistor did not seem to yield to this kind of approach. They could only be solved through a thorough understanding of the science involved.[50] Most engineers were not well equipped to deal with these new problems. They were in a strange position. The transistor was regarded as important in the early fifties largely because of its advantages as an amplifier over the traditional, established amplifier, the valve.[51] Their task was to eliminate the disadvantages of the transistor, but these had very little to do with valve technology. There was no merit in trying to make transistors familiar to engineers by introducing them, as was not uncommon, as a new sort of valve.[52] Transistors had markedly different circuit requirements: one could not simply pull out a valve and replace it with a transistor.[53] Nor was there much comfort in the superficial simplicity of the transistor, an apparent simplicity which had encouraged fascinating articles on how the point contact transistor might be made by the enthusiast in his own home.[54] The transistor was not a simple device and the problems posed in the manufacture and improvement of the device were especially complex.

> The transistor in 1949 didn't seem like anything very revolutionary to me. It just seemed like another one of those crummy jobs that required one hell of a lot of overtime and a lot of guff from my wife ... It wasn't exciting, not exciting at all. My job in the factory was to turn someone else's dream into saleable hardware. William Winter

To cope with these problems, were 'practicing electronic engineers who believe that the electronic principle, the control of free electrons in vacuum or gas-filled tubes, is an everlasting birthright and a sure touchstone of progress'.[55] Men whose whole experience was with valves, whose careers were built round a mastery of valve technology, could hardly be expected to be unreservedly enthusiastic about a rival newcomer which posed such strange problems. Nor could these men be expected to share the scientists' enthusiasm for a transistor which they did not understand, and which was so frequently justified by its superiority over the valve they understood. Some extreme observers even went so far as to suggest that the valve was rapidly on its way out,[56] and there was a more general, and surely galling, tendency to measure the transistor's progress in terms of the more sedate development of the valve over the previous forty years.[57]

About the enthusiasm of the scientific establishment there can be no doubt. In 1956, it went to the length of awarding the highest accolade available to it, the Nobel Prize for Physics, jointly to John Bardeen, Walter Brattain and William Shockley.[58]

Most electronics companies, whatever their enthusiasm for the transistor, felt it was a device that might acquire considerable importance and was, therefore, one in which it would be only wise to develop some ability. Research, and some development, was usually given to science-based research laboratories. But because the transistor was an amplifier, as was the valve, it was almost universal in the early days for transistor manufacture and development to be given to existing valve departments. In these, the transistor was often regarded, perhaps subconsciously, as the enemy, a threat to the status quo, and a challenge to personal, professional and corporate accomplishment.[59] Again, the transistor was thrown into conflict with the valve. While the new device might have a few specific advantages, the valve men could, with justification, scoff at the transistor's weaknesses. Most attempts to improve the transistor's reliability were, not unnaturally, based on established valve technology. In desperate bids to protect the transistor from the rigours of the outside world, engineers resorted to canning, hermetic sealing and glass encapsulation. Many early junction transistors were packaged in the type of can used for valves; the wires that had previously supported a valve, now held a semiconductor wafer in place.[60] It may have been a considerable comfort to many that the new upstart should be so heavily dependent on tried and proven techniques.

> The more enterprising people realised the opportunities and just jumped from one field into the other. There were many people who felt they had committed their entire career to vacuum tubes, magnetrons, klystrons and the like and they never changed.
>
> Frank Herman

> The people who first applied the transistor applied it as a substitute vacuum tube ... Basically it wasn't until you have a new generation of designers who don't have their antecedents in the vacuum tube and they can look at this from scratch and then you begin to optimise this, not as a substitute vacuum tube, but look at it on its own merits.
>
> Herbert Kleiman

> The original packing technology stemmed directly from the vacuum tube industry.   Quentin Kaiser

> The first assembly methods, the first high vacuum technology, the first actual commercial transistors of the junction type were packaged by means of the technology that had been developed for tubes.
>
> Maurice Apstein

> In Westinghouse, as in probably every major older company, semiconductor activity started in an applications sense ... in the division that was making the tubes for the same purpose ... It handicapped semiconductors because it made the transistor look like a replacement for the tube and it was a few years before people started to look and see what the transistor could really do in its own right.
>
> Gene Strull

It is not, perhaps, usually the role of scientists to escort their invention as far as the market place, but the difficulties of the early transistor were such as to require scientists, as well as engineers, to remain involved right up to the manufacturing stage. Not all scientists were happy in this new role.[60] Problems of unwanted and incomplete oxide layers on semiconductors, for example, could occasionally be avoided by the application of technological art – by 'cookbook recipes' – but they required the application of fairly basic science if they were to be permanently solved.

As late as 1953, there was even still a suspicion among some leading British physicists that the transistor was no more than a piece of good publicity for Bell.[61] Perhaps more typical of scientific attitude at this period was the feeling that if investigation of silicon and germanium had yielded such interesting applications, then it would be well worthwhile looking at the fundamental properties of all semiconductors. The transistor was the key that had opened the door to a whole new world of intriguing and intricate scientific investigation. The subject leapt from obscurity to such a pitch of interest that by 1956 an annual total of about a thousand papers was being produced on the properties of various semiconductors, including the complex semiconductor alloys.[62] Semiconductors, it was later claimed, had become the perfect vehicle for research rather than an end in themselves.[63]

Thus the transistor had something of a mixed reception ranging from wild enthusiasm to open hostility. The enthusiastic were largely those who did not appreciate the complexity of the problems. Neither, probably, did the hostile, but they were also unaware of the vast potential of the device. In the middle was a silent minority which realised how much work needed to be done and also had some inkling of the possible rewards. The transistor was not simply a new sort of amplifier, but the harbinger of an entirely new sort of electronics with the capacity not just to influence an industry or a scientific discipline, but to change a culture. The majority did not appreciate this because it looked upon the transistor in traditional terms, as only a new amplifier, and as such it was not, at first, impressive. What most failed to appreciate was the importance of the radically different way in which the transistor worked. This made the transistor more than just an invitation to study semiconductors in depth or a problem to be overcome to make a better valve. The transistor was to make feasible a radically new sort of electronics which would be capable of doing things previously impossible. That a radical invention should produce radical change is obvious only in the bright light of hindsight; at the time man must evaluate change in the light of his own knowledge and experience, useful tools when dealing with the familiar, but poor things with which to assess the value of the revolutionary.

> Everyone faces the future with their eyes firmly on the past and they don't see what's going to happen next.   J. R. Pierce

> You have to show an enormous advantage of something different to be able to get anybody convinced that you ought to use it rather than the existing technology and techniques . . . It really has to be useful in some application that you just can't meet with existing technology, and useful at a reasonable price.   Jerome Kraus

CHAPTER 6

A new industry

Science had presented the electronics industry with a new device in the transistor: it was then the responsibility of technology to produce the device in its most usable form. In attempting to do this, the industry encountered considerable difficulties throughout most of the 1950s.

> I think it's very important in this industry to differentiate between the ability to produce a new device in the laboratory and the ability to produce the thing in scale at competitive prices. Now the former was really much easier. The people who made money really did the difficult task which was producing these things in large quantities at very low prices. I think you have to look at that as two different problems. John Tilton

The germanium point contact transistor was in commercial production at Western Electric by 1951 and also in experimental production in some other large companies, such as Raytheon, at this time.[1] As the first commercial representative of the new electronics, the germanium point contact transistor was not an impressive creation. Certainly it marked a scientific breakthrough, but it was a manufacturer's nightmare. It was difficult to produce reliable transistors and even harder to make them with identical characteristics; life expectancy was uncertain and the transistors seemed inclined to deteriorate rapidly under temperature and humidity conditions which were far from extreme.[2] When these manufacturing difficulties were considered alongside the technical limitations of the point contact transistor, its inability to handle high frequencies and large currents and a tendency to produce a great deal of unwanted noise, the transistor seemed a troublesome and unattractive product.[3] It was even maliciously suggested that the point contact transistor made so much noise that at very low frequencies it should be capable of blowing itself up.[4] Fortunately, industry did not have to persevere with the point contact transistor for too long.

By April 1952, the junction transistor was in production at Western Electric, though at less than one hundred a month. Sample lots were also available from Raytheon, RCA and General Electric. In comparison, total monthly point contact production was about 8400; nearly all of which came from Western Electric.[5] At first, Shockley's junction transistor was made by pulling crystals from a melt of germanium. As the crystal was slowly withdrawn from the hot crucible, a doping pellet was added to the melt to give the crystal p-type conductivity. This was followed by another pellet to produce a thin layer of n-type and a third to produce a second area of p-type. Once the crystal was cut into slices across the junctions and leads painstakingly attached under the microscope to each of the three regions, a grown junction transistor resulted.[6] Such transistors produced less noise than point contact transistors, were more efficient and could handle more power, but suffered from the serious disadvantage of being restricted to even lower frequencies.[7]

The Bell Symposium of 1952 taught the art of crystal growing and the construction of transistors and no doubt encouraged commercial interest and activity, but an important development in the manufacture of junction transistors took place outside Bell in 1952, at General Electric. That company developed a method of alloying indium dots to opposite sides of a thin slice of germanium. The method produced an alloy junction transistor and was soon adopted by RCA and Raytheon as well as General Electric.[8] The alloy junction transistor could operate at higher frequencies and currents than the grown junction transistor, but only if the germanium layer were sufficiently thin. As final distance between the two indium pellets had to be no more than 0·01 mm, developing a sufficiently accurate manufacturing process proved difficult.[9]

A solution to this problem was provided by the ingenuity of the Philco Company in late 1953. The germanium wafer, it was discovered, could be eroded by spraying each side with an electrolyte while the germanium was at a positive potential relative to the spray nozzles. An infrared detector revealed precisely when the germanium wafer was of the required thickness and ready for the indium alloy.[10] The Philco jet-etching technology produced better quality transistors and in the surface barrier transistor achieved the highest frequency response yet attained. This transistor was always expensive and the thinness of the germanium base rendered it rather delicate.[11]

Of much greater importance, considering the future course of semiconductor electronics, was the announcement by Texas Instruments in May 1954 that they had succeeded in making a silicon transistor. The accomplishment was announced by the man most responsible, Gordon Teal, to the Institute of Radio Engineers and directly followed other papers which had declared that a silicon transistor would be impossible for many years.[12] But while other companies had tried to use the most recent manufacturing methods to produce silicon transistors, Texas Instruments, on Teal's advice, had explored the older grown junction technique and had been successful. The silicon transistor, with its capacity to work in much higher temperatures than any germanium transistor, was of much greater interest to the Military.[13] This was to become a highly significant factor.

Another new manufacturing technique, which proved of major importance to the future direction of electronics, was a diffusion process to make first germanium and later silicon transistors. Work at Bell and at General Electric led to a process by which an impurity was allowed to diffuse from the vapour phase into the semiconductor. By regulating the time allowed and the temperature at which the process took place, the penetration of the impurity into the semiconductor could be controlled with accuracy. The addition of sophisticated photographic techniques permitted the laying of intricate mask patterns on the semiconductor so that diffusion took place only in the desired areas. Masking by silicon dioxide was eventually to prove particularly successful. Not only did this new sort of transistor achieve yet better frequency performance, it was also more reliable. More important, the diffusion process meant that for part of the production process, transistors could be treated in batches rather than individually. So significant did the process seem that Bell held a second symposium in 1956 to inform the industry. Germanium diffused transistors were in commercial production at Western Electric by that time.[14] Various refinements were soon made to transistors produced by the diffusion process and it could be argued that some at least were major innovations in themselves.[15] It is impossible to look at all the variations. While there had been sixty types of transistors with appreciably different characteristics in 1953,[16] by 1957 there were no fewer than six hundred.[17]

As we have seen, the discovery of the point contact transistor and later that of the junction transistor were very much triumphs not only of physics, but of chemistry, metallurgy and of electrical engineering.[18] The importance of materials technology cannot be overstressed. By late 1948, Gordon Teal and J. B. Little of Bell Laboratories had managed to produce single crystals of germanium by slowly pulling the crystal from the melt. While this overcame the considerable problems presented by polycrystalline germanium, it did not solve that of accurately doping the germanium with impurities. That was only accomplished in 1950, again at Bell and again by Teal – this time working with Morgan Sparks – when the crystal pulling apparatus was adapted to allow the addition of impurity pellets to the melt in the way that produced the grown junction transistor.[19] Until the production of this transistor, Teal encountered considerable opposition at Bell from those most concerned with transistors, including metallurgists, who thought that single crystals were unnecessary and that their production would be much too expensive a process for commercial transistor manufacture.[20]

The advent of the junction transistor, in which impurity levels and distribution were critical, probably encouraged more materials work at Bell to supplement the success of Teal. Between 1951 and 1954, W. G. Pfann developed the process of zone refining by which a narrow molten zone was swept through a horizontal rod of germanium. As impurities in the germanium have a greater affinity for molten rather than solid germanium, the impurities tended to be carried along to the end of the rod where they could conveniently be cut off. The process could be repeated several times and yielded greater purification than was possible by any other method. Zone refining was not so remarkable as a discovery – the principle had been used for decades to refine aluminium[21] – as for the degree of precision Pfann's adaptation of the principle allowed. Modern high frequency heaters helped. When required, levels of impurity could be reduced to one part in a thousand million to produce material of the highest purity then known.[22] Yet further adaptation of the process was possible and Pfann developed it to grow single crystals horizontally to supplant the awkward method of pulling them vertically from the melt. It was also possible to use the apparatus as a zone leveller, that is to ensure the equal distribution of impurities throughout the crystal. So successful and relatively simple were these techniques that they were rapidly and universally adopted by the industry.[23]

Making pure silicon was a much more difficult business. Silicon has a much higher melting point than germanium and a strong tendency to absorb impurities from any crucible containing it. Large single crystals of silicon had been grown at Bell by 1951, but purity remained elusive. In 1953, a floating zone refining technique was developed at Bell by which surface tension separated the hot silicon from its container, permitting the production of reasonably pure material. But most materials research into silicon, including that at Texas Instruments, whence the first commercial silicon transistor was to emerge, concentrated on the chemical purification of silicon. This had been the direction of research taken by the University of Pennsylvania during the war when it had collaborated with Du Pont in a programme to produce pure silicon.[24]

Despite the very valuable ability of silicon transistors to operate at high temperatures, various obstacles delayed the use of silicon for transistors and not the least of these was the difficulty of making the silicon sufficiently pure. While silicon's energy gap of 1.2 eV, compared with that of 0.72 for germanium, permitted higher temperature working, it also demanded that the space between electrodes be smaller. Obtaining close enough spacing for germanium was difficult enough under manufacturing conditions. But even had the problems of material purity and electrode spacing been overcome, the silicon transistor, because of the lower carrier mobility of silicon, suffered inherently from a frequency performance that was inferior to that of the germanium transistor.[25]

Consequently, in the period up to 1955, nearly all transistors were made of germanium rather than silicon.[26] As late as 1958 it was incorrectly estimated that 75 per cent of all transistors would be made of germanium until well into the sixties.[27] While a large percentage of the earth's crust is composed of silicon in the form of sand, germanium is a rather rare element. At the time of the invention of the transistor, annual world production of germanium was about six kilograms, obviously too little to support any sort of transistor industry.[28] Fortunately, germanium was available as a by-product of other extraction processes. In the United States, it could be obtained from zinc refining in the form of germanium dioxide. In this form it cost about $300 per kilogramme in 1952: by the time it had been purified it was worth more than gold by weight, though the germanium in each transistor weighed only about 0.002 gramme.[29] In Britain, germanium was present in certain coal seams and its concentration was much greater in the dust which collected in flues when the coal was burnt. Various chemical processes could be used to extract the germanium and reduce it to a suitable form, and possible production of germanium from British coal was actually very much greater than that from American zinc.[30]

Both silicon and germanium are in group IV of the periodic chart; that is, each molecule has four valence electrons in the outer ring. As both elements presented serious obstacles to their use in transistors, it seemed that it would be possible to create compound semiconductors – such as the copper oxide or zinc oxide used in devices in the thirties – by combining elements from group III with those from group V, to give an average of four valence electrons. Indeed, this certainly was possible and semiconductor diodes were constructed of such compounds as indium antimonide, and gallium arsenide. Such intermetallic compounds provided a theoretical solution to the limited frequency response and vulnerability to high temperatures of group IV semiconductors, but problems associated with achieving an exact purity balance between the elements were daunting.[31] The very limited use made of III–V compounds and later of II–VI compounds has never justified the sums spent on their research. Despite predictions in the late fifties that the future of transistors lay with such compounds,[32] they have been of comparatively little practical use. The future of semiconductor electronics lay in quite another direction. A paper of 1963 highlighted the problem. At that date, the best developed of the compound semiconductors was gallium arsenide and it probably still is. But while a developmental silicon transistor might cost $60 and drop to $6 in quantity production, a gallium arsenide transistor would start at $600 and drop to $60 with a mass market. Where would be the market for such a device at such a price?[33] Germanium and silicon had an unassailable lead.

> Until Gordon Teal grew single crystals of silicon, everybody was on germanium which turned out not to be the best way to go. If the same amount of effort had been put into silicon carbide or gallium arsenide or other II–VI and more exotic compounds, we might not be so deeply involved in the materials we are using right now. But they got so far ahead, you see, that for the others to catch up, the inter-metallics and so forth, would have taken so long.
>
> George Abraham

In 1951, there were four companies in the United States believed to be making some transistors commercially. By 1952, there were eight; in 1953, fifteen; and by 1956 no less than twenty-six.[34] In the early fifties, General Electric, RCA and Sylvania were the three largest receiving valve manufacturers in the United States, accounting for about three quarters of total production. The only other valve manufacturers of significance were Raytheon, Philco, CBS (Columbia Broadcasting System), Tung-Sol and Westinghouse. By 1953, all eight of these companies had become transistor manufacturers. The remaining seven companies were Western Electric, the manufacturing arm of the Bell System, and six new companies. In this context, the term 'new' is applied to any company starting to make transistors without previous experience in electronics. A firm such as Germanium Products, founded solely to make transistors for hearing aids, is an example of a new company, but so too would be the large consumer products firm, Motorola, which was early to start limited transistor production for its own use, or Texas Instruments, a small geophysical services company which saw a future in the new electronics and commenced transistor manufacture in 1953.[35] Many more new companies were to enter the semiconductor business as the fifties progressed and it is widely held that the entry to the industry of so many new firms has been directly responsible for much of the highly individual flavour of the industry and indirectly responsible for the massive impact on society semiconductor electronics rapidly came to have.

A cursory examination of some of the major transistor innovations during the first half of the fifties has already suggested that the established receiving tube firms, along with Bell, played a dominant role in improving transistor design, performance and manufacturing processes. As late as 1959, this group of eight firms and Bell were contributing 57 per cent of the total R & D expenditure on semiconductors and 63 per cent of major innovations, most of the latter coming from Bell.[36] Table 6.1 shows semiconductor patent awards to these groups and suggests the same pattern of contribution to the development of the new product.

Understandably, the contribution of Bell Laboratories declined from the near monopoly of knowledge it held at the beginning of the fifties. The contribution of the new firms increased as they began to enter and understand the semiconductor business, but the major contribution of the established valve companies at this period can be in no doubt.

Table 6.1. *Percentage of total semiconductor patents awarded to firms in the U.S., 1952–6*

	1952	1953	1954	1955	1956
Bell Laboratories	56	51	46	37	26
Receiving tube firms	37	40	38	42	54
New firms	7	9	16	21	20

Source: John E. Tilton, *International Diffusion of Technology,*[34] p. 57.

An important difference between the valve firms and the new firms at this period was the marketing aggression of the latter. Table 6.2 shows the percentage of the total semiconductor market held by the leading firms in 1957.

Table 6.2. *Percentage of total U.S. semiconductor market, 1957*

Western Electric		5
Receiving valve firms		
General Electric	9	
RCA	6	
Raytheon	5	
Sylvania	4	
Philco-Ford	3	
Westinghouse	2	
Others	2	
Subtotal		31
New firms		
Texas Instruments	20	
Transitron	12	
Hughes	11	
Others	21	
Subtotal		64
Total		100

Source: John E. Tilton, *International Diffusion of Technology,*[34] p. 66.

Despite all the money and effort put into R & D and the fine record of patent awards and innovation of the valve companies, the semiconductor market was early and rapidly falling into the hands of the new companies. This obviously requires some explanation. Between 1954 and 1956 there were some 17 million germanium transistors and about 11 million silicon transistors sold in the United States, altogether worth about $55 million. During the same period, over 1300 million receiving valves were sold; a market worth over $1000 million. Sales for all electronic components for those years amounted to about $6500 million.[37] Transistors and other semiconductor devices, while regarded as of interest and possibly of long-term importance to the established electronic companies, were hardly likely to be their main concern. Receiving valves, still regarded as competitors of the transistor, had a growing market, one which reached its peak in numbers produced in 1955, in value in 1957, and which did not decline seriously until the second half of the sixties.[38] Even then, average price per valve did not decline. Indeed, it has shown a steady, if gradual, increase since the fifties and, as the percentage of special purpose valves grows, continues to do so.[39] Hence, the established electronics firms were certainly not in the position of having to abandon a stagnant or decaying traditional electronics industry for the rigour and risk of the new field. It is even argued that the exhaustive research which went into semiconductors helped stimulate new valve work at a scientific rather than a technical level with such revolutionary results as the klystron travelling wave tube.[40] Traditional electronics in the fifties was expanding and it was a huge market compared with the tiny transistor market.

Raytheon was one of the established valve firms of the early fifties, but exceptional in that it became more involved with transistors than the others. Having begun production in 1951, by 1952 Raytheon was producing more transistors for the open market than any other company.[41] By March 1953, Raytheon was producing 10 000 junction transistors a month at about $9 each, exclusively for the hearing aid market.[42] This seemed a natural market for the early transistor as not only did it make use of the transistor's most obvious advantages, but it was also a market that could be exploited despite the early limited supply of transistors and their high cost.[43] But, having dominated one market, Raytheon never managed to extend its transistor production to others. By 1957, Raytheon still controlled 80 per cent of the small hearing aid transistor market, but had fallen well behind in other types of transistors.[44] By 1960, despite its early lead in transistors, Raytheon was no longer among the top semiconductor companies.[45]

In contrast, Texas Instruments started to take an interest in semiconductors in 1949, but had had no previous concern with the manufacturing of electronic components.[46] From 1930 until 1950, the firm was known as Geophysical Services Inc. and carried out oil exploration surveys under contract. The move to electronic instruments, occasioned by a desire to avoid the vagaries of survey work, coincided roughly with the invention of the transistor.[47] Texas Instruments noted the development and approached Bell for a licence in 1951, but it was not until after the Bell symposium of 1952 that Texas Instruments put any effort at all into semiconductor research and development.[48] The research laboratory was not established until January 1953 when Gordon Teal left Bell to head it. From that time, Texas Instruments committed a great deal of money, for a small company, to three distinct projects. The first of these was to make a silicon transistor by the grown junction method. This became commercially available in the spring of 1954. The second was to make a mass market device using germanium transistors. The first commercial transistor radio – the Regency – was produced in October 1954 in collaboration with the IDEA Corporation. The third project was to produce large quantities of pure silicon by chemical methods. This was accomplished by December 1956.[49] Some of the explanation for this extraordinary pace compared with that of Raytheon lies in the emphasis Texas Instruments puts on specific research goals and on innovation in production and marketing as well as in research. The silicon transistor, for example, was positively rushed from research to production to marketing so that its problems could be shared all along the line, and so the purists in research would not lose time by seeking perfection.[50] The resulting silicon transistor, with its appeal to the Military, gave Texas Instruments a highly profitable three year monopoly in the industry,[51] and it can now be seen as one of the earliest indications of the importance commercial considerations were to assume.

The activities of a third company – Transitron – during this early period are also interesting as an example of yet another approach to the new industry. Transitron was founded in August 1952 by two brothers, Leo and David Bakalar. Leo had been a businessman in the plastics industry, David a solid state physicist at Bell. Their factories in Massachusetts were a converted bakery and, later, an old knitting mill. The company's name suggests the product the Bakalars had intended making, but in 1952 the transistor market was almost entirely the hearing aid market and that had already been captured by Raytheon. So the new company turned to a simpler semiconductor device, a point contact germanium diode. Work had recently been going on at Bell on a diode with a welded gold whisker, and it was just such a device that Transitron offered for sale in 1953.

The gold bonded diode was capable of handling much higher voltages than other diodes and was consequently rapidly accepted by the Military. Transitron was always heavily dependent on the military market and, for many years, on its one original product. Exactly how original this was, given Bell's work on the device, may have caused some comment in the industry, but the important fact is that Transitron was successful in manufacturing the device effectively. Transitron put very little money or effort into research. Yet the company made money. Despite its somewhat dilapidated factories, its lack of interest in research and innovation or even new markets, Transitron became the second largest semiconductor manufacturer by the mid fifties and perhaps the most profitable by the late fifties. Transitron lacked the calculated aggressive drive of Texas Instruments and even the initial burst of innocent enthusiasm shown by Raytheon, yet it succeeded.[52] Why was it that a company with such apparent disadvantages as Transitron should be so successful in the semiconductor business?

Part of the explanation lies in the nature of the innovations at the heart of the new industry. Despite the early interest in the transistor as a better valve, the transistor was so radically different from the valve in the way it worked, in the way it could be manufactured and sold, and in its apparent potential, that it could not be comfortably accommodated within the existing electronics industry without changes that that industry was then unwilling or unable to make. In its typical subjugation of semiconductor development, manufacture and marketing within valve departments, the established electronics industry demonstrated that it was largely unaware of the impact the innovation could have. Even where this awareness existed in the industry, it was often not accompanied by an appreciation of the new pace of change and of the entirely new problems which rapid change presents to industry. But even had there been a thorough realisation of the significance of semiconductor electronics and of the speed with which developments would occur, the large electronics firms might still have felt they were doing all they could. After all, their

record of semiconductor innovations and patents and of money
spent on semiconductor research and development during most of
the fifties far surpasses that of any other group. What they failed to
realise was that this would not be enough to secure mastery in the
new field, that the secrets of success lay with individuals rather than
corporations and that supremacy would fall to those who could attract, keep and make best use of these individuals.

Bell Laboratories had long been regarded by many as a sort of
post-doctoral training school and it was quite normal for good
young scientists to spend two or three years there before moving on
to other jobs. With its virtual monopoly of transistor knowledge in
the late forties and early fifties, and the leading role it played in
semiconductor development throughout most of the fifties, the
Laboratories became an exclusive school for semiconductor scientists and technologists. Not surprisingly, Bell did its utmost to attract good scientists who had already worked on suitable subjects. A
typical example might be Alan Goss who had finished his Ph.D. on
crystal growing at Southampton in 1952. Goss sent his thesis to Bell
and was later interviewed by William Shockley and specifically
questioned on the importance of controlling purity in germanium
crystals. Goss was recruited and worked under Pfann on materials
problems for two and a half years before returning to England.[53]
Such an experience was fairly typical. To a lesser degree, the large
electronics firms with their considerable semiconductor research
and development efforts also acted as training establishments for
those who sought to learn about the innovation. That scientists
should have left Bell with newly acquired semiconductor knowledge
can hardly have been too disturbing for Bell. After all, the diffusion
of semiconductor knowledge was the very purpose of the symposia
of 1952 and 1956. But the flow of trained manpower from the old
valve companies, whose sole task was to compete successfully in the
open market, must have given cause for grave concern.

The scientist or technologist working on semiconductors within
one of the large electronics corporations at this time rapidly found
himself in an unusual position. He became the repository of
knowledge that was scarce and valuable. Yet, within the established
electronics company he commanded the prestige and salary of any
other employee of similar years and standing. As his semiconductor
department was merely a unit of the valve division and subject to the
priorities and policies of the whole electronics organisation, he

often felt under-valued and constrained. Semiconductor research and development in the early fifties was a highly scientific endeavour and it did not fit comfortably into the settled technological atmosphere of traditional electronics, nor was it likely to be best directed by an established management, totally unused to this type of work and whose main concern lay not with product innovation, but with cost reduction.[54] The transition in semiconductors from laboratory curiosity to factory product was neither smooth nor easy. It demanded a scientific and technical expertise that was in short supply. In 1951, the Telecommunications Research Establishment in England had sought to employ anyone who knew what the word 'transistor' meant,[55] and not until that date did the first American university – Harvard – begin a course in semiconductors. Consequently, those who possessed semiconductor knowledge were an elite group in a field that was then wide open. By 1955 there were about a dozen American universities with solid state courses, and a typical graduate could expect five or ten job offers.[56] In contrast, the universities were producing nuclear physicists in droves and many of these had to adapt to the demands of the time. New ways of making better materials, production improvements, new devices and new applications for them were occurring almost weekly in the early fifties and were frequently the result of individual effort or that of small groups. It was hardly surprising that these individuals should seek to use their knowledge as efficiently and as profitably as they could.

> Most of the people who were key people in the development of the transistor did not have their training in solid state physics. In fact, the single largest group involved in the development of the transistor were nuclear physicists, by far. They certainly outnumbered engineers and it turns out that other groups, like biologists, were quite important. Harvey Brooks

> In the early days, top management did not appreciate the uniqueness of those relatively few engineers and scientists who understood semiconductor devices. Moreover, they frequently put electronic engineer managers in charge of their semiconductor operations that didn't know a thing about the chemistry and metallurgy of semiconductor devices . . . For about a year, we had a going-away party every Friday. Clare Thornton

Examples of men who left companies to use their precious semiconductor knowledge elsewhere are not hard to find. Gordon

Teal and David Bakalar have already been mentioned. William Shockley himself left Bell in 1954 to start his own semiconductor company.[57] Transitron's first employee, Gunther Rudenberg, came from the semiconductor group at Raytheon and the second from General Electric.[58] There were few others. One of them, who had previously worked for RCA and Raytheon, remembers there being only six or seven qualified engineers or scientists in the old bakery Transitron used as a factory.[59]

No doubt many skilled men left one company for another in the hope that they would make more money in the new company, but it seems that very many found working in the large electronics firms frustrating. They disliked the remoteness of bureaucratic decision making and the established hierarchy, but most of all, they resented what they saw as the lackadaisical attitude towards semiconductor work. Those who knew nothing about semiconductors might regard them only as materials with some potential, but those who understood them knew very well what the potential was and were positively straining at the leash to play a part in realising it. They had suddenly, and perhaps unexpectedly, found themselves at the frontier of knowledge and were anxious to make their mark before others advanced that frontier beyond their reach. They required only the opportunity and if that were not forthcoming within one company, they were willing to go elsewhere to find it. Philco, one of the few old firms to have a separate semiconductor group in the early fifties, were so anxious to retain one young scientist with a brand new Ph.D. in solid state physics that, though he had had only six months' industrial experience, they took the unusual step of offering him an astronomic salary increase and a holiday in Europe.[60] Another semiconductor man who worked for this company for three years during this period felt that working for Philco did not carry sufficient prestige. Attracted by the personal reputation of Shockley, he left Philco to work for him in 1954.[61]

> When we joined Transitron, we really joined out of plain scientific interest. We did not know that from an old bakery we could create respect from companies like Fairchild, like TI, who consider us one of their main threats.   U.S. electronics expert

Part of the explanation for the strength of this feeling of personal involvement and participation rests with the methods that had to be used to make semiconductor devices in the early fifties. There was no existing technology. Everything had to be invented or improvised or adapted and then altered as new developments changed requirements. Every company was striving to find new technical processes to gain an edge over its competitors. One man, once an aeronautical engineer, still takes pride in the crystal puller he built at this time, apparently far superior to anything made at Bell.[62] As the new field required contributions from many disciplines, so men with any expertise that might be relevant were recruited. At Raytheon, the semiconductor division employed a block drilling engineer from the automotive industry and a golf ball expert.[63] There were no semiconductor service industries at this time. Raytheon had to purchase germanium dioxide procured from zinc mining companies, reduce it to germanium, pre-refine it, zone refine it, make single crystals and so on. All the crystal furnaces were made by the company and were manually operated.[64]

Highly scientific the heart of a semiconductor device may have been, but many of the methods used to produce these devices in the early fifties were far from scientific. The effects of slight variations in furnace temperature and other factors that had an influence on production yield were learned the hard way, by experience, and such knowledge was obviously valuable. Methods employed at this period have been described as involving art rather than science,[65] as witchcraft[66] or black magic.[67] In fact, the methods of some companies were embarrassingly empirical. To make semiconductor diodes, one company used to melt rather impure silicon in a quartz crucible, add brass, cool the whole lot slowly, break it apart, remove the quartz, saw the polycrystalline silicon into lumps, attach leads, hit them until the device showed desirable electrical characteristics, and then screw the leads fast.[68]

Despite the rapid technological innovation throughout the early fifties, semiconductor manufacture was generally dependent on rudimentary and *ad hoc* principles. It was common to make all semiconductor components on just one basic assembly line. Usable devices were selected from the wide variety produced on the line, their functions decided by their electrical characteristics and their price by the scarcity of the component.[69] The difficulties of making transistors, of encapsulating them, and of controlling the impurities in the semiconductor material continually troubled the industry and were responsible for the exceptionally low yields that were universal. Problems were so severe, that a mythical element which killed transistors – 'deathnium' – was postulated; this was eventually found to be copper.[70] Extreme empirical measures were often resorted to, such as 'forming' the point contact transistor by putting a high current through the emitter electrode. In the early fifties, the assembly methods for transistors were commonly no more sophisticated than lines of women using tweezers to manipulate components under microscopes. An unintentional variety of products resulted from these manufacturing processes. A typical batch of alloy transistors might be found to contain 20 per cent high-frequency units selling at $6 each in 1960, 20 per cent lesser frequency units selling at $1.50 or $2.50 each, 40 per cent entertainment quality transistors each worth between 45 cents and $1.50, and 20 per cent worth nothing.[71] Such problems of quality variation within a batch of supposedly identical transistors were largely overcome by new production methods which became available within the industry in the early sixties. Before this, manufacturers had often to sell whatever they could produce rather than produce what they could best sell.

> The thing that put semiconductors aside from any other technology that I know of up to that time, were the very low yields. In the mid to late fifties yields of 20 to 30 per cent were common and even lower on some sophisticated devices. Gene Strull

The first commercial devices made of transistors produced by these methods were hearing aids, but transistors were generally too expensive, too limited in their possible applications and too unreliable to be extensively used in commercial applications in the early fifties.[72] Of more importance at this period were semiconductor diodes and rectifiers, components which had benefited considerably from semiconductor research and development, particularly from the work on materials. Germanium was increasingly used to make diodes for use in drive motors, welding equipment, battery chargers, amplifying circuits, radios and television sets.[73] As less sophisticated versions of these components had been in use for some time and as the latest semiconductor devices were fairly compatible with traditional circuit design, the growing use of such components aroused little comment. This was not the case with the transistor. Because its potential was often recognised to be great, it was sometimes assumed that its dramatic and widespread use in commercial applications was imminent. For example, because Bell had started to use transistors in a card translator[74] and experimentally to generate signal tones in dial switching equipment in one small exchange in 1952,[75] it was assumed that the transistor would make long distance dialling possible and could be used as a hardy amplifier in underwater telephone cables.[76] In fact, Bell was not to risk putting transistors in submarine cables until 1966,[77] and grave reservations began to be expressed about whether the transistor would be as reliable as existing, perfectly serviceable equipment or even compatible with it.[78] An indication of the sort of problems involved might be the fact that Bell were still producing germanium point contact transistors for card translators long after such transistors were obsolete, because their replacement with newer transistors would have meant re-designing the whole card translator system.[79]

Similarly, the application of the transistor in television receivers, though frequently heralded at this time,[80] was to be another fifteen years in the realisation.[81] Again, transistors for use in automobile electronics, in fuel ignition systems and automatic headlight dimmers were imagined to be imminent,[82] but these developments have taken even longer. The reasons for this frustrated anticipation are complex and will be dealt with later, but one is relevant now. When the transistor was first announced, it was looked upon as a valve. Appreciation of its potential and of the difficulties presented by its commercial manufacture had earned the transistor recognition in its own right by the early and mid fifties, but with the result that successful commercial production of a functional, reliable, efficient and cheap transistor was regarded as an end in itself. The device was seen as a radical improvement on anything that had gone before and it was not envisaged that the device would itself undergo radical improvement. The main difficulties were production difficulties and these were seen to include such matters as transistor type and performance. Better transistors were produced because there was a demand for the improved article, and also because the new article was often easier to make than its predecessor. Of course, the best way of all to make vast quantities of cheap, reliable and identical goods was seen to be the automatic assembly line. This seemed particularly appropriate for transistors as it was neatly imagined that transistors would actually control the very machines that were manufacturing other transistors.[83] It was also in stark contrast to that laborious manual assembly of transistors that was the reality of the early fifties. Logically, if that were the main obstacle to commercial transistor production, its opposite – automatic assembly – could be viewed as the best solution.[84] It was not. Change in the new industry was to proceed at such a rate as to preclude this method of manufacture for many years. It is this very rate of change that has been in part responsible for the transistor finding rather different markets than those which were originally envisaged for it.

The first commercial transistor radio used Texas Instruments germanium transistors. The following year, in 1955, Raytheon brought out its own model selling for $80.[85] Car radios were being built with transistors shortly afterwards,[86] and, though television receivers presented difficulties, that was not true of television cameras.[87] One of the most attractive, and immediate, though rather limited, uses of transistors was in early computers. Computers of the forties, such as the ENIAC, had required thousands of valves. The disadvantages were obvious; such machines occupied a massive volume and so did their power supply. The valves generated a great deal of heat which had to be dispersed; and they also needed continual replacement. The transistor was made attractive for use in computers by the disadvantages of valves.[88] IBM marketed an early commercial computer in 1955 which replaced 1250 hot valves with 2200 transistors. Size was reduced, the need for cooling removed, and power consumption reduced by 95 per cent.[89] The computer provides perhaps the best example of the transistor replacing the valve because of the weakness of the valve rather than any advantage of the transistor. This can even be seen in the general attitude towards the computer at the time. Babbage's idea for a computer in the 1830s had been technically impossible. The valve changed that and computers became practical though with a very limited role. The transistor, it was imagined, would make computers better machines, but it was hardly considered that it would make them able to play a vastly greater role. For example, a Department of Commerce survey of the late forties on potential usage of computers concluded that large – by forties' standards – digital computers would only be used by such organisations as the Census Bureau and that perhaps one hundred would satisfy the nation's entire need.[90] Change, it seems, is most readily contemplated one step at a time.

Table 6.3 gives an indication of the vigorous growth experienced by the whole electronics industry during these years. The Government had naturally been the biggest purchaser during the war, but was rapidly toppled by the post-war consumer spending spree. The

Table 6.3. *Uncorrected value of U.S. electronics industry sales by end-use category, 1950–6*

	Total sales ($M)	Government (per cent)	Industrial (per cent)	Consumer (per cent)	Replacement components (per cent)
1950	2705	24.0	13.0	55.5	7.5
1951	3313	36.0	13.6	42.3	8.1
1952	5210	59.5	9.5	25.0	6.0
1953	5600	57.7	10.7	25.0	6.6
1954	5620	55.2	11.5	25.0	8.3
1955	6107	54.6	12.2	24.6	8.6
1956	6715	53.5	14.2	23.8	8.5

Source: EIA, *Market Data Book,*[103] p. 2.

Korean War once again brought the military market well to the fore, a position it was to retain for many years. By 1955, the Government was buying 22 per cent of all semiconductor devices made, and accounted for 35 per cent of the value of the semiconductor market.[91] It is well worthwhile trying to assess the effect on semiconductor development of such a huge military interest.

In no sense had the invention of the transistor been a military achievement. Bell had been anxious from the beginning to steer clear of military involvement and did not receive their first R & D grant from that source until 1949.[92] The Military was treated much like one of the electronics companies and, like them, it had to assess for itself the value of the transistor. Like the companies, the Military saw the transistor as a new sort of valve. By 1952, the Department of Defense had established a Subpanel on Semiconductor Devices within its Research and Development Board, subservient, of course, to the Panel on Electron Tubes. Earliest concern seems to have been for the most obvious applications; the possibility of a wrist watch radio, for example.[93] Of the 90 000 transistors produced in 1952, mostly point contact devices from Western Electric, the Military bought nearly all.[94] Like the purchasers of the new hearing aids, the Military in the early fifties was not primarily concerned with price. Instead, it was more concerned with such matters as availability, whether performance could match military specifications, and whether semiconductor devices would be reliable. It was also necessary for military electronic engineers to develop new circuits suitable for the new devices. One considerable attraction of the new devices was the reduction in weight they offered, both in the components themselves and in the batteries or power supplies needed to power them.

The Military's growth of interest in electronics had been a fairly recent development. A destroyer of 1937 carried only sixty valves: the destroyer of 1952 carried 3200. As a study of this latter year revealed that 60 per cent of Naval electronic equipment in the fleet was not operating satisfactorily, even this modest amount of electronics was obviously giving cause for concern.[95] As half of all failures were caused by valve problems, a device which promised to replace the valve was obviously an attractive proposition, whatever the cost. Anyway, costs of even traditional military electronics had soared and promised to continue to do so. While a wartime bombsight had cost $2500, the latest computing bombsight of 1952 cost $250 000.[96] In that year, the Air Force estimated that about 40 per cent of its electronics could be handled by transistors, with a saving of 20 per cent in size, 25 per cent in weight and – incredibly optimistic – 40 per cent fewer failures.[97] A year later, the Military began to appreciate – as did industry – some of the problems associated with making and using transistors. A report of 1953 read, 'The transistor has been a quite unreliable device during this project. Of the one hundred type M1698 units used during development only about twenty-five remain within specifications and few of these have the same characteristics as when received . . . It is disconcerting to place a unit in a negative resistance display test instrument and watch as the curve jumps from good to bad and back; or to move the unit in the socket and watch the characteristics change'.[98] Obviously the Military needed semiconductor electronics, but it was going to have to put a great deal of work into development before it could derive the benefits it expected.

In 1951, the three Services assigned responsibility for military production development of the transistor to the Signal Corps. The role of these production engineering measures (PEMs) was to increase the availability of the transistor, reduce its cost and improve its performance and reliability. Most Government money devoted to semiconductors was distributed in this form rather than as R & D contracts with industry. For example, the Signal Corps distributed less than $500 000 annually in the form of contracts for specific semiconductor R & D work before 1956, and about $1 million per year after this date. PEM support is estimated to have totalled about $50 million between 1952 and 1964.[99] In 1955, total Government R & D funding to the semiconductor industry amounted to $3.2 million, while production refinement funding for transistors and diodes was about $4.9 million.[100] One aim of the military emphasis on production was to ensure adequate production capacity. A condition of production funding was that the receiving company should make itself able to produce 3000 units of each component per month.[101] It is possible that such a policy on the part of the Government in fact led to a degree of over-capacity. In 1954, the industry had sold 1¼ million transistors: by 1955 it had the ability to produce 15 million annually,[102] and actually made only 3.6 million.[103]

Not surprisingly, the first firms to benefit from Government money were the large, established electronics companies. The first Signal Corps PEM contract went to Western Electric in 1952, and the four other companies supported later that year were General Electric, Raytheon, Sylvania and RCA.[104] In 1956, the Signal Corps embarked on a huge PEM programme for semiconductors, devoting almost $15 million to the project, nearly all to transistor production. By this date, some of the new entries to the industry, such as Transitron, Motorola and Texas Instruments had attracted support.[105] Yet, by 1959, Western Electric and the eight valve firms were still receiving 78 per cent of Government R & D funds though they held only 37 per cent of the total market. New firms, with only 22 per cent of the Government's R & D funds, had, by then, captured 63 per cent of the semiconductor market. At this time, the older companies were themselves providing about two-thirds of their R & D expenditure and the new firms about 90 per cent.[106] It is imagined, though not known, that this situation was also typical of the early and mid fifties. The semiconductor industry as a whole benefited from Government R & D expenditure and more from Government production improvement funds, but the older firms benefited very much more than the newer.

Where both old and new firms appreciated the Government most was in its role as a market for semiconductor devices. We have already seen that defence requirements accounted for about a third of the value of all semiconductor sales in the mid fifties. By the late fifties, that figure had grown to about a half.[107] For some companies, such a market was vital. Transitron was utterly dependent on it for some years after 1953 and the value of the silicon transistor to Texas

Instruments was largely in the military market. Military support for new firms by the provision of a market seems to have been essential to those firms. By 1959, 69 per cent of the military semiconductor market had been captured by such firms, an even greater percentage than their share of total semiconductor sales.[108] Such a market was particularly valuable because its eagerness for the latest and best components, and its limited concern with price, meant that new and expensive components could be sold at a time when production costs rendered such components too expensive for the industrial or commercial markets. Until experience and greater production brought down such costs, a phenomenon referred to as a 'learning curve', the Military was likely to be the only large market for semiconductor components. In the early fifties, when alternative customers were scarce and all semiconductor components expensive, the role of the Military as a buyer was particularly crucial. In this industry, with its complex and critical dependence on production and price, this was vital.

> I can hardly think of a single company [in the U.S.] in the fifties that did not enjoy significant government support for their semiconductor operation ... I can absolutely assure you that we would never have enjoyed the success that we enjoyed in that period had we not had the government money.   Clare Thornton

> At Texas Instruments our approach was to take advantage of the military money anyplace we could, but that we were going to build products that were sellable in other places.
>   Bob Cook, formerly of Texas Instruments

CHAPTER 7

The pace of progress

The direction of semiconductor development during the early and mid fifties was determined not so much by the desire to make better devices as by the desire to find better ways of making them. Certainly there was interest in improving characteristics, in creating new types and in extending the application of semiconductor devices generally, but primary interest was in improving the processes by which devices could be made in quantity. The better product was important, but more important was the ability to make devices that were reliable, reproducible and cheap. One of the major innovations of the fifties, perhaps the major innovation, was the planar technique, producing what was in some ways an inferior transistor reliably and cheaply. It is the power of process over product which has determined so much of the course of semiconductor electronics.

The planar process was firmly based on the diffusion and oxide masking techniques which had been developed at Bell and General Electric and which had been made available to the rest of the industry by means of a second Bell symposium, in 1956. The diffusion process was a vast improvement over the older alloy or grown junction techniques in that the depth of impurity diffusion into the semiconductor could be accurately controlled by adjusting furnace temperature and the time the semiconductor was exposed to an atmosphere of gaseous impurities. Hence, the thickness of the base layer – a critical factor controlling the frequency performance of the device – could be adjusted more than ten times more accurately than previous processes allowed.[1]

The first product of this new precision was the mesa transistor, so named because it looked like a mesa mountain in miniature. The mesa was produced by diffusing an outside layer of p-type material into a wafer of n-type germanium or silicon. A thinner layer of n-type was then added to the outside of the p-layer and three sides of the wafer subsequently cut away. Large junctions between n- and p-type material then remained, too large to give satisfactory performance, and these were reduced by masking a central portion with a protective oxide and etching away the peripheral, unprotected areas. The oxide was then removed and electrical contacts made to the raised structure – the 'mesa' – to produce a transistor.[2]

The mesa transistor had many advantages in addition to the ease with which the thickness of its base region could be regulated. It worked in the high-frequency range, it was rugged, it dissipated heat readily and, because a degree of batch production was possible during much of its construction, it was cheap to produce.[3] There were, however, the usual serious disadvantages. Electrical connections still had to be made manually and this was both slow and expensive. Moreover, though it was quite feasible to create a base region that was only 0.005 mm thick, the region had to be at least ten times thicker to permit handling without breakage. Also, once the oxide was removed from the mesa surface, that surface was liable to contamination and no encapsulation seemed capable of ensuring permanent protection.[4] But the mesa was undoubtedly a vast improvement over anything that had gone before and by 1960 it seemed likely that the process would become universal and perhaps even permanent.[5] This was not to be the case.

The planar process, which has dominated semiconductors since its inception, was the triumph of Fairchild Semiconductor. The company was started by eight scientists who sought and achieved the financial backing of the Fairchild Camera and Instrument Corporation in 1957 to allow them to attempt further exploitation of the relatively new diffusion and oxide masking techniques. The eight had all previously worked at Shockley Semiconductor Laboratories where effort had been directed towards discovering new devices such as a four-layer diode.[6] Having been attracted to Shockley's company by his scientific reputation, the dissident scientists felt that such heavy concentration on the basic scientific aspects of semiconductors would be less rewarding than paying attention to that area where opportunities were greatest – process technology. The founding of Fairchild Semiconductor was a result of the power of commercial motivation over scientific and the consequence was planar technology, the technique by which nearly all modern semiconductor devices are manufactured.

The planar process is the key to the whole of semiconductor work.
G. W. A. Dummer

The planar technique was similar to that used for the production of mesa transistors, but it had some crucial advantages. Bell had thought in terms of pure silicon dioxide as a mask and this had been inadequate to prevent the diffusion of such impurities as phosphorus through the mask to the supposedly protected silicon beneath. Fairchild found that when the oxide mask itself contained phosphorus, passivation of lower layers was complete.[7] Moreover, if this passivating layer of oxide were retained on the surface of the device, protection of the delicate junction was greatly improved. As with the mesa transistor, the planar technique allowed much of the fabrication to be accomplished from one side of the wafer, making for production economies. Unlike the mesa, the planar transistor, as the name implies, was flat. This meant that electrical connections did not have to be made laboriously by hand, but could be achieved by depositing an evaporated metal film on appropriate portions of the wafer. Limitations of the planar technique were that it was suited only to low power devices and to silicon rather than germanium, for a good oxide layer is more easily produced on silicon and metal films are compatible only with silicon oxide.[8]

The planar technique depends upon the repetition of three basic processes. First, oxidation is used to create a mask over the wafer; next, 'windows' are opened by a photographic method in selected parts of the oxide layer; finally, impurities are diffused into the exposed silicon. The sequence is then repeated so that diffusion layers of impurities can be implanted exactly where required. This is achieved by means of photo-resist techniques – originally and incorrectly called photolithography[9] – by which a plan is cut out on a large sheet of opaque film, in which the transparent regions will eventually determine the 'windows' in the silicon wafer. A step and repeat camera is used to gradually and accurately reduce the pattern to the size of the wafer and to repeat the pattern to form many identical devices on the same wafer.[10] Improved photo-resist techniques meant that planar became even more successful as a batch production technique than the mesa process had been, particularly as batch production could be extended to the connection stage. The use of the photo-resist technique, a variant of photolithography, is yet another example of interdependence of technologies and of cross-fertilisation. The method had been developed for printing purposes and had been in use in this area for some time.[11] It is but one out-standing example of the adoption and adaptation of extraneous technologies to improve the manufacture and design of electronic devices.

The planar process, devised in 1958 and in commercial production by 1959, though not announced until the following year,[12] was improved by a Bell innovation of 1960. Epitaxy was a means by which silicon vapour could be deposited on a single crystal silicon substrate. Components could be formed in the deposited silicon by planar diffusion without interfering with the substrate. Hence the substrate contributed mechanical strength to the device without also giving the undesirable electrical characteristics of a thick base region.[13]

The second half of the fifties and the first few years of the sixties was the period during which the semiconductor industry found its feet and the self-assurance to grow into a major industry, not just within electronics, but in its own right. At the heart of this confidence and activity were realistic manufacturing processes, such as diffusion and planar, which could be used to produce components with predetermined characteristics reliably and cheaply. The *ad hoc* manufacturing technology of the early fifties, while sufficient to sustain – at least with military help – a minor industry, would have been unable to support the sort of growth the industry was experiencing by the early sixties. Much of this growth was not only permitted by the new processes but actually precipitated by them, for batch production in general and planar in particular, prompted both a rapid increase in the numbers of components produced and an even more rapid decline in their price. The older manufacturing methods produced devices which allowed the market to become familiar with semiconductors, but which in no sense dominated that market. The new methods produced devices which did dominate the market previously prepared.

> The development of the transistor on a very large scale depended, as I see it, on that step of getting away from germanium, getting away from having to protect your surface artificially with layers of wax or grease, or worse still, putting glass round it.   D. A. Wright

In the decade from the mid fifties to the mid sixties the number of American companies producing transistors doubled while the number making diodes and rectifiers more than tripled (see Table

7.1). The total quantity of such units manufactured soared and so too did the value of this production (see Table 7.2). An integral part

Table 7.1. *American companies making* (1) *transistors,* (2) *diodes and rectifiers*

	Transistors			Diodes and rectifiers		
	Number of firms	Entries	Departures	Number of firms	Entries	Departures
1957	22	1	2	26	—	—
1958	24	2	0	38	12	0
1959	27	4	1	57	21	2
1960	33	6	0	59	6	4
1961	37	6	2	74	17	2
1962	41	7	3	81	12	5
1963	40	6	7	84	14	11
1964	34	1	7	85	10	9
1965	43	9	0	85	3	3

Source: Jerome Kraus, 'An economic study of the U.S. semiconductor industry',[6] pp. 118, 120.

Table 7.2. *Production and uncorrected value of U.S. semiconductor components*

	Transistors		Diodes and rectifiers	
	Units (M)	Value ($M)	Units (M)	Value ($M)
1957	27.8	68.0	66.0	83.0
1958	46.1	112.0	74.8	98.0
1959	83.5	228.5	129.4	166.5
1960	131.8	313.9	197.6	228.0
1961	193.0	316.2	312.0	248.6
1962	258.2	303.5	424.3	267.7
1963	302.9	311.7	627.1	282.3
1964	398.0	323.1	762.1	312.0
1965	631.7	425.9	1072.6	379.4

Source: Jerome Kraus, 'An economic study of the U.S. semiconductor industry',[6] p. 80.

of this growth was the new importance of silicon following the invention of the silicon transistor, but especially associated with the growing use of the planar technique after about 1960. Even more significant was the decline in price of the average semiconductor component during these years (see Table 7.3), caused by technical

Table 7.3. *U.S. sales of germanium and silicon transistors (uncorrected values)*

	Germanium		Silicon	
	Units (M)	Average value ($)	Units (M)	Average value ($)
1957	27.7	1.85	1.0	17.81
1958	45.0	1.79	2.1	15.57
1959	77.5	1.96	4.8	14.53
1960	119.1	1.70	8.8	11.27
1961	177.9	1.14	13.0	7.48
1962	213.7	0.82	26.6	4.39
1963	249.4	0.69	50.6	2.65
1964	288.8	0.57	118.1	1.46
1965	333.6	0.50	274.5	0.86

Source: EIA, *Market Data Book*,[37] p. 87.

advance and by fierce competition. A silicon transistor of 1957, for example, would typically have cost nearly $20, while that of 1965 would probably have cost less than $1. Yet average prices mask the even steeper decline in price of individual components. A specific transistor produced by Texas Instruments sold for $6.75 in 1959, but only three years later for $1.13, while a particular Fairchild transistor was priced at $19.75 in 1959 and only $1.80 in 1962.[14] Consequently, various attempts to measure the productivity of the industry during this period have come to diverse conclusions. When productivity is measured by value added per employee, it appears that the American semiconductor industry increased its productivity by a mere 8.8 per cent between 1958 and 1963, well below the average for the rest of American industry. Measured in terms of shipments per employee, though, the increase was a startling 253 per cent over the same period.[15]

> The hungry firm is primarily responsible for driving the prices down to a point where they are sometimes unrealistic for anyone to stay in the business. This is true in any industry.   William Winter

One of the problems with the semiconductor business is that it has moved too fast from a business point of view. It's never realised the potential economic gains it could make ... Nobody in this business seems to have learned from past experience. William Corak

As the total production of transistors had accelerated, so too had the variety of transistors manufactured (see Figs. 7.1 and 7.2). Between 1956 and 1962 some 6000 different transistor types had been produced, of which about three-quarters were still available in

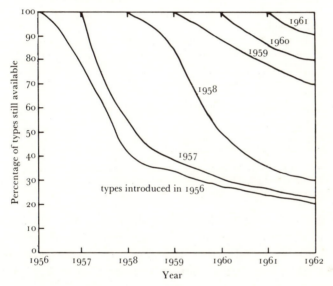

Fig. 7.1. Soure: Jerry Eimbinder, 'Transistor industry growth patterns',[16] p. 58–9

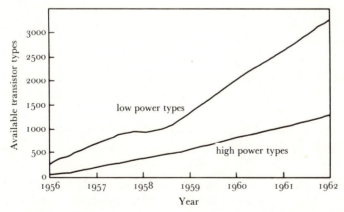

Fig. 7.2. Source: Jerry Eimbinder, 'Transistor industry growth patterns',[16] p. 58–9

1962.[16] The depression of the early sixties marked a definite turning point. Before 1963, the variety of transistors available increased by 48 per cent each year: since then the rate has dropped to about 18 per cent.[17] Electrical characteristics of transistors were steadily improved. In particular, new transistors were devised to handle higher and higher frequencies (see Table 7.4). Transistor frequency perfor-

Table 7.4. *High-frequency transistors commercially available*

Maximum frequency (MHz)	Number of available types			
	1959	1960	1961	1962
400–599	2	11	39	52
600–799	2	9	17	19
800–999	1	1	8	8
1000–1199	0	1	1	6
1200–1999	0	0	3	5
2000–3000	0	0	0	2

Source: Jerry Eimbinder, 'Transistor industry growth patterns',[16] p. 58.

mance improved at such a pace that William Shockley felt it possible in 1958 to construe a frequency production index by dividing the cut-off frequency value for the highest frequency transistor produced during a year by the volume of transistor production during that year.[18] The suggestion that improvement in transistor performance kept pace with the volume of transistor production is welcome and probably justified, at least up to 1960, but the implication that there is some mathematical relation between these two factors is unjustified.

The massive growth in semiconductor production during this period was leading to an increased diversification of the market for its products. Adoption of semiconductor devices by users demanded either the replacement of conventional electronics in existing systems, or the manufacture of entirely new systems in which semiconductors could be incorporated to advantage. The latter was much more important in promoting the use of semiconductors: the former accelerated as a consequence of the success of semiconductors in new spheres. One such sphere was computers, largely and increasingly dependent on semiconductor electronics. In 1955, only

150 computer systems were sold in the United States, making a total of only 250 then in use. In 1965, 7400 computer systems were produced and 31 000 were in use by that date.[19] Table 7.5 gives some

Table 7.5. *Value of U.S. transistors by usage, 1963 ($M)*

Military		Industrial		Consumer	
Space	33.0	Computers	41.6	Car radios	20.6
Aircraft	22.8	Communications	16.0	Portable	
Missiles	20.3	Test and		radios	12.6
Communications	16.8	measuring	11.7	Organs and	
Surface systems	10.8	Controls	11.5	hearing aids	7.3
Strategic systems	8.8	Other	11.5	Television	0.3
Other	6.7				
	119.2		92.3		40.8

Source: Edward Ney Dodson III, 'Component product flows in the electronics industry',[38] pp. 95–7.

idea of the importance of computers in the transistor market in 1963. Computers alone accounted for about one-sixth of transistors bought in that year. Indeed, the computer market was then as big as the whole consumer market, itself overwhelmingly dominated by the use of transistors in radios. The main purchaser of semiconductor equipment was the Government. Table 7.6 gives some indication of the amount of semiconductor production absorbed by the Military between 1955 and 1965. These are minimum figures given by manufacturers when they knew their products were bound for a military destination. It is likely that greater proportions ultimately found their way into military use. Clearly, the military market was of tremendous significance, reaching a peak of importance in about 1960, when nearly half the value of all semiconductor shipments was consumed by the Military, though it has been suggested that if figures were available for the period 1952–4 they would show an equally high military consumption.[20] The regularly higher proportion of military purchases by value than by quantity is explained by military demand for the newest components and for those which would meet rigorous specifications. The greatest demand came from the Air Force and from NASA, both particularly concerned with reliability and with reducing size and weight. The Navy and Army were perhaps less eager and needed to be persuaded of the advantages of each new device.[21]

Table 7.6. *Percentage of U.S. semiconductor production designated for military use*

	By quantity	By value
1955	22.0	35.0
1956	22.2	35.6
1957	21.3	35.8
1958	21.5	38.5
1959	25.8	45.5
1960	28.6	47.7
1961	25.0	39.3
1962	22.0	38.4
1963	17.0	33.0
1964	14.9	24.7
1965	11.8	23.6

Source: Jerome Kraus, 'An economic study of the U.S. semiconductor industry',[6] p. 80.

Encouraged by the huge military market and by the growing demand for semiconductor components in new sorts of electronic systems, the new companies of the late fifties, with their new batch production methods, precipitated an enthusiasm for semiconductor electronics that for the first time seized many more than merely those who understood how such devices worked. By the late fifties, semiconductor devices were beginning to appear as commercial products in their own right. Commercial enthusiasm for the product was shown in the large number of new companies which entered the business and in the excitement these companies generated in financial circles. For example, shares in Texas Instruments, which had been priced at $5 in 1952, had soared to $191 by 1959.[22] Transitron sales jumped from $1 million in 1954 to $42 million by 1959 and its profit after tax from $0.07 million to $6 million during the same period.[23] Until 1959, Transitron remained a private company, but late that year the Bakalar brothers sold 13 per cent of the company's stock at a price which implied a market value for the firm of some $285 million.[24] For a completely new firm of only seven years' standing this was success indeed, but it was only an extreme example of a feeling that semiconductors were important and were going to become very much more important.

Glamour industry though semiconductors may have been in 1959 and 1960, a rude shock awaited in 1961 and hard times persisted until 1963. Tables 7.2 and 7.3 give an indication of the problem. Production grew steadily during these years, but was not matched by a comparable increase in the value of sales. Production, in other words, had outstripped demand and the consequence was severe price competition. This does not mean that there was necessarily a reduction in demand; simply that its increase failed to keep pace with accelerating semiconductor production. The sudden semiconductor depression of the early sixties occurred in the midst of the national recession of 1961–3,[25] but there appear to have been other causes. The considerable profits which made the semiconductor business such a glamour industry in the late fifties had attracted the participation of a large number of competing companies. So too had the industry's rapidly changing technology, which put the newcomer at no disadvantage to the established firm. Table 8.4 (p. 117) shows the concentration of semiconductor shipments by company size and reveals some renewed surrender of a portion of the market by the larger firms to the smaller between 1958 and 1962. This share was not recaptured by the bigger companies until the mid sixties.

As the industry began to settle down to a commercial pattern, the technologist-managers, so successful in the fifties, were beginning to show weaknesses. Components were made and sold with very little regard to the cost of manufacture, while buyers purchased semiconductor components with a primary regard for price and only the most secondary concern for quality. This made the sudden and significant intervention of the Japanese in the American semiconductor market all the more worrying. In 1959, the Japanese who had previously played virtually no part in the American market, started shipping considerable quantities of cheap germanium transistors, generally for use in transistor radios (Table 7.7).

> Whereas most industries are strongly influenced by marketing considerations, the semiconductor industry is still basically run by technically-trained people who think they can pull off anything . . . The thirteen years I spent in the semiconductor industry I never saw a marketing forecast that came true . . . They are really technical people who figure if they can make it, somebody will buy it, rather than marketing people who figure there's a need and I'm going to fill it.
>
> U.S. semiconductor executive

Table 7.7. *Japanese semiconductor exports to the United States*

	1956	1957	1958	1959	1960	1961
Transistors						
Units (thousands)	—	1	11	2400	3400	5700
Total uncorrected value ($M)	4	1	7	1.6	1.8	2.1
Transistor radios						
Units (thousands)	—	—	—	4000	4200	4600
Total uncorrected value ($M)	—	—	—	57	55	48

Source: Arthur D. Little, *Patterns and Problems of Technical Innovation in American Industry, Part IV,*[26] p. 147.

In 1961, the relative importance of the Military as a customer began to decline for the first time in many years. As the Military was the industry's largest market, often virtually its only market for expensive devices and, for many companies, a principal support, this may have been responsible for some lack of confidence. Of more definite importance, though, were the various measures the Military had taken throughout the fifties to increase semiconductor production potential. By the early sixties, when over-production was seriously eroding prices and profits, the only defence of many firms was to make use of the extra capacity enforced by the Military and so compensate for low prices by making yet more components.[26]

Figure 7.3 shows the annual percentage reductions of prices of transistors and of diodes and rectifiers between 1955 and 1965 and gives some idea of the severity of the crisis between 1961 and 1963. Price decline had, of course, been a prominent feature of the semiconductor business since its inception, and was to continue long after the financial maelstrom of the early sixties. Indeed, it is still one of the most outstanding characteristics of the industry.

> Semiconductor products are the only class of products that historically have been cheaper in any subsequent year.
>
> Dean Toombs

> But most managers in semiconductors really believe they can make something at any given day for half the cost that their accountants tell them to make it for and their net result is to fire the accountant and get a new one ... I grew up in it. I grew up in situations where I

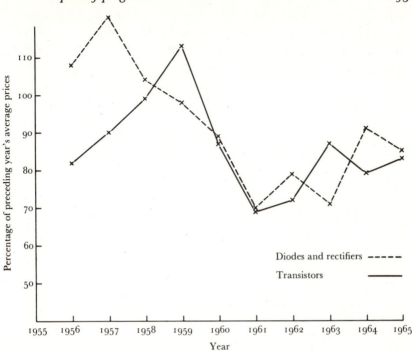

Fig. 7.3. Average uncorrected prices of transistors and of diodes and rectifiers as a percentage of preceding year's prices. (After Jerome Kraus, 'An economic study of the U.S. semiconductor industry', [6] p. 32)

> figured out I could make something for $8 and my boss said you got-ta make it for 6 and I said I can't make it for 6 and I saw behind his eyes that if you can't make it for 6 I'll go hire someone else who will make it for 6. Then I realised the other guy was just a better liar. So I said all right and I'll figure how to make it for 6. I made it for 6. By then some other company was offering it for 5 because they had a bigger liar.　　　　　　　　　　　　　U.S. semiconductor executive

The learning curve concept is one that is of use throughout manufacturing industry, but which is particularly relevant to the semiconductor industry. It depicts the relationship between volume of production and unit cost, by which unit cost is seen to decline as production volume grows.[27] For example, Texas Instruments has found that every time production volume doubles, the price drops to 73 per cent of its previous level.[28] Because of the continual and sometimes precipitous fall in prices of semiconductor components, which has made steep learning curves a feature of the industry, the learning curve concept has often been used to explain some of the more extreme consequences of the pace of semiconductor progress.[29] Figure 7.4 is a model, based on learning curve theory, of what typically happens to a semiconductor product. The component is first produced in limited numbers and sold at a higher price,

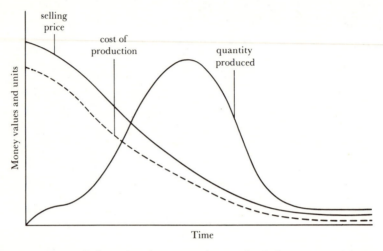

Fig. 7.4. The typical semiconductor product cycle. (After A.M. Golding, 'The semiconductor industry in Britain and the United States',[8] p. 92)

though perhaps at a loss. As production quantity increases, the price drops sharply and also draws closer to the cost of production curve. Cost of production eventually stabilises and the price stabilises just above it. But this often happens only when the device is obsolescent and reduction in demand has caused a major and permanent decline in the quantity produced. At this point, small profit margins and an unpromising future would normally result in withdrawal from the market or the substitution of an improved product. While this model is applicable to most industries, most do not suffer its extremes. In contrast, the semiconductor industry commonly undergoes the whole cycle for each component type in a very few years. Moreover, there is usually a vast difference between initial demand and peak demand and, as we have seen, between starting price and ultimate price. Table 7.8 gives the relevant figures for production and sale of one type of transistor from one company between 1959 and 1962. Clearly, the product was approaching the peak of its quantity curve and had reached a stage in which price was so low that even increased production would soon no longer yield a satisfactory profit.

Table 7.8. *Sample learning curve for one company's transistor type*

Item	1959	1960	1961	1962
Components sold (thousands)	400	2900	7300	12200
Total Cost ($)	3700	7152	12802	12710
Cost per component ($)	9.25	2.47	1.75	1.04
Price ($)	6.75	3.46	2.10	1.13
Revenue ($)	2700	10034	15330	13786
Business profit ($)	−1000	+2882	+2528	+1076

Source: William Long II, 'Price and nonprice practices',[39] p. 60.

As a user, I would much prefer a product stabilised at a higher price, knowing that I'm always going to be able to buy that product and get delivery. I'd rather do that than watch the price go down to 30 cents and then worry my head off that I can't get the parts.

Richard Gerdes, Optical Electronics Inc.

Running a business in this commercial climate was not the easiest of tasks and it was rendered yet more difficult by demands made upon the industry for forward pricing. A company had not simply to sell whatever it made for whatever it could get. Rather it had to cater for the long-term demands of customers who were anxious to order in advance and to pay future low prices rather than current high ones. Consequently, the semiconductor company had to forecast its rate of price reduction for some years ahead and this naturally left room for error. Table 7.9 shows the forward pricing of one semiconductor firm for a specific component.

Table 7.9. *Forward pricing ($)*

Year for which forward price applied	Year in which forward price given			
	1959	1960	1961	1962
1960	2.70	—	—	—
1961	1.60	1.70	—	—
1962	1.25	1.20	0.90	—
1963	1.10	1.05	0.70	0.65
1964		0.92	0.61	0.50
1965			0.58	0.44
1966			0.55	0.42
1967				0.41

Source: William Long II, 'Price and nonprice practices',[39] p. 78.

Yield – the proportion of total units to emerge from all manufacturing procedures as working units – improved steadily with the
adoption of more efficient manufacturing processes. The processes
of the early fifties often produced a yield of no more than 10 per
cent or 20 per cent. At the end of the decade, yields of between 60
per cent and 90 per cent were typical of the mesa process.[30] More
important, though, were the huge yield increases made when an individual company became fully conversant with its own variation of
a manufacturing process. For example, in May 1960 Fairchild was
producing an advanced silicon transistor at a net yield of between 1
per cent and 3 per cent of started units. The young company,
anxious to meet orders and encourage more, boosted production.
At the same time, a change in etching technique so improved yields
that not only were all orders immediately met, but 200 employees
were suddenly thrown out of work.[31]

Such erratic production meant that the industry was often faced
with what were known as inventory problems. Unavoidable or even
accidental production of components with characteristics which
rendered them not immediately saleable meant that stocks of these
components could easily grow to alarming proportions. In the
seller's market that prevailed throughout most of the fifties, demand
was sufficiently high to accommodate a still limited and somewhat
erratic supply of semiconductor components. By the late fifties,
though demand had increased considerably, the ability to produce
semiconductor components in quantity had developed at a greater
pace. Consequently, inventories of components for which there was
least immediate demand filled manufacturers' warehouses. There
was little else manufacturers could do with such components and
there was always the chance that they would fill some future order.
Yet, product change was so rapid in the industry that any products
consigned to a warehouse were likely to become obsolete before a
buyer could be found. Not surprisingly, it was tricky assigning a
value to such stock. Was it to be valued at cost of manufacture or at
market price? In fact, the technical management of the semiconductor industry was seldom aware of the niceties of accountancy and
such surplus stock was often valued very much more highly than it
should have been. Only when hard times came in the early sixties
and it became unavoidably clear that much of this stock was
worthless, did many companies face up to reality. In June 1961, for
example, Transitron reduced the estimated value of its inventory

stock by $7 million at a stroke, a move which was the main factor in turning a 1960 profit of $8.1 million into a 1961 loss of $1.5 million.[32]

The semiconductor depression of the early sixties caught everyone unawares. Not only was the industry unprepared for this reverse, it actually foresaw a period of unprecedented opportunity and prosperity. Manufacturers had come to accept continually improving process technology, rapid product change, declining prices and the participation of a large number of firms as an integral part of a distinctive and distinguished industry. By 1960, semiconductors had come to epitomise progress through technology and science and the semiconductor industry provided the means of effecting this progress. The industry had made possible the technological exploitation of science and that had brought progress. In 1960 it was inconceivable that any factors would arise to halt the march of progress.[33] Floor space devoted to semiconductor research, development and production in 60 companies rose from 3.3 million square feet in December 1958 to 4.9 million in April 1960, but the companies then estimated that this latter figure would double by April 1962.[34] An individual firm provides a fairly typical example of the dynamism of the times. In 1956, Motorola commanded 8 per cent of the transistor market and was aiming at 20 per cent. By 1957, the company planned to have increased its payroll from 380 to over 500.[35] In 1960, though, plans were even more ambitious. Despite declining prices, Motorola that year tripled its 1959 sales, more than doubled its floor space and planned to increase its workforce from 2000 to 3200 by mid-1961. A double shift was in operation on the mesa transistor line producing 75 000 transistors a day and plans were afoot to work a triple shift.[36]

The reversal suffered by the semiconductor industry in the early sixties was a timely warning that things could not go on as they were. No matter how clever or useful the device, more and more companies could not have continued for ever producing greater and greater quantities at ever decreasing prices. The industry of the early sixties was badly in need of a new sort of product, better suited to its now sophisticated production methods than were discrete components. Demand for semiconductor products was evident and future demand seemed certain to be much greater, but increased production on the sort of scale and at the sort of prices that seemed likely suggested only commercial disaster. If the demand for semiconductor electronics were to be satisfied and the scope of semiconductor electronics realised, then the industry patently had to do very much more than simply look to its production methods. It had to produce the semiconductor product in a new form.

CHAPTER 8

The integrated circuit

The integrated circuit was, among other things, a solution to the commercial bottleneck problem – too many companies producing too many discrete components at too low prices – in which the semiconductor industry found itself by the sixties. The integrated circuit was a commercial innovation developed by scientists working in a technological industry. Thus, it contrasts with the transistor, which was a scientific invention discovered by scientists who had little connection with industry. Throughout the fifties, this basic scientific invention was gradually transformed into a commercial product by means of the development of industrial technology. The integrated circuit was a product of the further development of this technology for commercial purposes.

> I have always had an interest in trying to change things or advance the state of the art in one way or another and my interest in this was primarily because I thought that it was a scheme that could dramatically impact the way electronic equipment was built. I guess that was enough motivation.   Jack Kilby, Texas Instruments

An integrated circuit is one in which the functions of several discrete components are performed within a single piece of semiconductor material.[1] For an industry whose concern was the production of vast quantities of separate components which would eventually be linked in circuits to perform useful functions, it was a logical step to supply complete circuits. The industry's latest batch production technology pointed in the same direction. If it were possible and economic to make hundreds of identical components on one wafer and then to cut the wafer in order to sell separate components, then surely it would be feasible to make a suitable selection of components on the wafer, to connect them in the same way that connections were made to form circuits from individual transistors, and to sell whole integrated circuits.[2] In fact, this was basically how integrated circuits did come to be produced, though both the actual method of manufacture and the evolution process leading up to integrated circuitry are vastly more complicated.

The patent for the first integrated circuit was filed in February 1959 by Jack Kilby of Texas Instruments.[3] His integrated circuit was, in fact, two circuits constructed in one piece of germanium. Despite its strength in silicon technology, Texas Instruments had, like Bell over a decade before, chosen to work in germanium because that was the better understood material.[4] Kilby had joined Texas Instruments in 1958 from Centralab where he had been working on silk screen techniques for printing a substrate onto which germanium transistors could be soldered to provide a complete circuit function. It was work which provided Kilby with an awareness of the desirability of integration, but which itself was limited in scope. The move to Texas Instruments was made because Texas Instruments' broader interests in the same area seemed to offer more opportunity for success, but definitely not specifically to work towards integrated circuitry.[5] Success came quickly and Kilby's first germanium integrated circuit had been made by October 1958. In fact, like the first point contact transistors, it was something of a monstrosity and would have been a production nightmare. Each separate transistor, diode and resistor within the integrated circuit had to be precisely and laboriously interconnected by hand.[6] Unknown to Kilby, this obstacle was already being overcome at Fairchild with the development of a batch production method which allowed metal connections to be evaporated onto the semiconductor surface. The method was, of course, the planar process.

> We were in a position at Fairchild where we had, because of the particular way we had decided to make transistors by leaving an oxide layer on top of them, one more element than everybody else did to build into the package to make the integrated circuit, and that happened to be the one that worked.   Robert Noyce, Fairchild

> The planar technique took the infant integrated circuit concept from the laboratory and made it feasible for production.
>
> Herbert Kleiman

So important was it to have an easy means of making connections between the various parts of an integrated circuit that a debate raged for years between Kilby at Texas Instruments, who had shown that separate components could be contained within one chip of semiconductor, and Robert Noyce at Fairchild who had demonstrated some six months later the process by which these components could be easily and economically connected.[7] The planar process also had the advantage of its insulating surface oxide layer which conveniently isolated each area of the semiconductor. As this layer was attainable with silicon and not with germanium, the planar process did much to initiate the dominance of silicon and to change completely the previous situation in which silicon was used only for expensive devices and germanium for cheaper ones. Without planar techniques, integrated circuits were technically just possible: with planar, they became commercially viable. The semiconductor industry of the early sixties knew very well the importance of that particular difference.

As with the invention of the transistor, it is argued that the creation of integrated circuitry occurred when the time was ripe. If the announcements had been delayed until 1960, it is felt that many more people than Noyce and Kilby would have been involved in patent disputes.[8] Certainly there was a demand for the new technique both from the industry and from its market, but there was also a long history in the electronics field of thinking in this direction. Sometimes the thinking had been vague and often it had been directed towards an end to which integration was no more than a possible means, but then the sources of invention are seldom clear and obvious. As the transistor had been the product of much diverse scientific effort over a considerable period, so the integrated circuit was the product of technological effort over a decade of semiconductor device manufacture. Both the transistor and the integrated circuit were products of their time, but the transistor emerged from the science of its age while the integrated circuit came from the technology of a later age.

> In contrast to the invention of the transistor, this was an invention with relatively few scientific implications . . . Certainly in those years, by and large, you could say that it contributed very little to scientific thought. Jack Kilby, Texas Instruments

Over decades, one prevailing concern of the electronics industry was miniaturisation. Reductions in size, weight and power consumption were important, particularly to the Military; and the Military was an important customer of the electronics industry. But associated with these reductions, and probably one of the most enticing reasons to make electronics smaller, was the feeling that

Fig. 8.1. Valve, transistor and integrated circuit. (Courtesy of Bell Laboratories)

miniaturisation would increase reliability.[9] Widespread though this conviction was, the greater reliability of small semiconductor components remained unproven even by the late fifties.[10] The very obvious advantages of smaller size seem to have been sufficient incentive, but the possibility of greater reliability at no greater cost made doubly sure that the miniaturisation movement was self-sustaining.[11]

In the same innocent way that the standard of work in the industry at this time was commonly measured by the percentage of staff with doctorates in a company,[12] so the accomplishments of the industry were measured in a similarly *ad hoc* and less than logical way. Achievement was calculated in packing densities, what the industry now recalls – with some embarrassment – as the 'numbers game'.[13] In 1950, 1000 tubes could be fitted into a cubic foot, though they might have melted under these conditions; in 1956, 10 000 transistors into the same space; and, by 1958, 1 000 000 components were to be squeezed into the same cubic foot (see Fig. 8.1). Future rate of progress was to be measured by how rapidly even greater contraction could be brought about.[14] The measure of packing density may not have been an entirely logical measure of excellence, but it was simple and was widely used.

Throughout the forties and fifties, improvement in valve technology was generally losing momentum except in one field. There was a demand from the Military and from certain other specialist users for smaller valves, for what were known as miniature valves and, eventually, for subminiature valves. Not only were such valves smaller, but they were also a better, more reliable though more expensive product because the customer demanded it and because the industry could devote a large proportion of its technological skills to this innovating sector. Hence, there was certainly justification in the traditional electronics industry for associating reduced size with improved product. The use of printed circuit boards for mounting valves and passive components started the trend towards circuit modules as integral units.

By the mid fifties, the transistor and other semiconductor components were becoming increasingly acceptable and familiar in applications, such as hearing aids, portable radios and airborne electronics, for which small size was a primary requirement. Of course, other advantages of semiconductor electronics, such as low power requirements, potential ruggedness and long life, were apparent, but never quite so apparent as the single attribute of small size (see Fig. 8.2).[15] The first commercial use of the integrated circuit, in December 1963, was, predictably, in hearing aids.[16] Small size was generally, though tacitly, used as an umbrella term to cover all the real and anticipated advantages of semiconductor electronics. There was, inevitably, resulting confusion about whether small size was an attribute of improved electronics or improved electronics an attribute of small size. When the electronics industry began to make semiconductors a commercial reality, it was by means of production methods that were able to take advantage of the tiny dimensions possible with such components as transistors. Commercial production processes were developed to supply small components and were doing this so successfully by the late fifties and early sixties that the development of new processes to produce larger components, even had these been electronically superior, would have been unacceptable.[17]

> People were enamoured with miniaturisation. That was the key word, how tiny it could be . . . It caught the imagination of the public and of the Military as well . . . if we made one mistake – business mistake – we should have emphasised size more. We did not.
>
> Gene Strull, Westinghouse

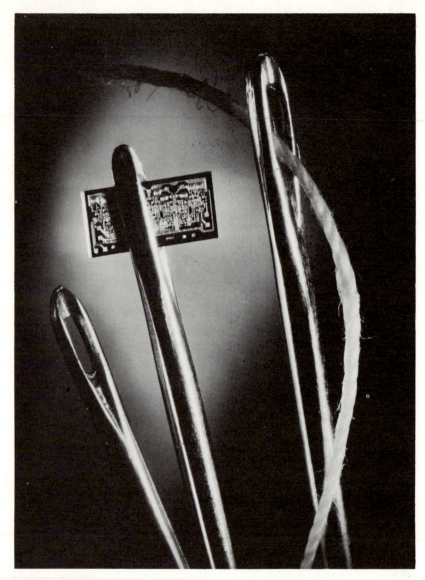

Fig. 8.2. Integrated circuit containing 120 components passing through the eye of a needle. (Courtesy of Mullard Limited)

The Tinkertoy project is perhaps the best example of early and continuing effort towards miniaturisation. Started as a Navy project in 1950 by the National Bureau of Standards, the idea was to produce smaller, automatically-assembled electronic systems. To this end, silk screen printing techniques were used to form passive components on ceramic wafers 22 mm square and $1\frac{1}{2}$ mm thick. Centralab, where Kilby was working, was one company involved in the development.[18] The printed circuits on the wafers connected the components to riser wires leading to other wafers mounted above and eventually to active elements – valves – on top. Almost $5 million had been spent on Tinkertoy by 1953, but the project had then been outpaced by events. The disclosure of the Tinkertoy project in September of that year revealed that the whole scheme was founded on valves and that no account had been taken of the invention of the transistor more than five years previously.[19] Years later, the project was revived in the Army Signal Corps' micromodule plan of 1957. Again, the aim was the possibility of greater automation, but this time transistors were to be used, the whole stack was to be permanently encapsulated and miniaturisation was to be a primary rather than a secondary goal.[20] By 1963, $26 million had been poured into the programme, much of it into RCA, and it was anticipated that by 1964, a million units a year would be produced.[21] In fact, that did not happen and the whole project ended as ignominiously as its predecessor. As Tinkertoy had been outpaced by developments, specifically the transistor, so the micromodule programme became a dinosaur in the sixties as it was overtaken by more successful miniaturisation techniques.[22]

Another approach to the problem of miniaturisation was the creation of what were known as 'microcircuits' on thin films. Various passive components were deposited on a thin ceramic substrate in which active semiconductor components were later embedded. Deposited metal strips made electrical connections to all components and the whole wafer, generally about 12 mm square and 0.75 mm thick, formed a complete two-dimensional circuit as opposed to the three-dimensional tiers of wafers used in micromodule schemes.[23] Much work on microcircuits had gone on at the Diamond Ordnance Fuze Laboratories, a military research organisation with an established interest in compact and reliable electronics. The microcircuit programme there started in 1957, the Laboratories succeeding in eliminating individual component encapsulation and in using printed circuit techniques to achieve what it called its 2D circuit.[24] Those responsible for much of this work at the Laboratories trace a great deal of their inspiration to close links with Bell Laboratories in the early and mid fifties, and particularly to research going on at Bell into diffusion and oxide masking. This, they say, enormously increased the utility of the D.O.F.L.'s knowledge of photoresist techniques. The step and repeat camera, and the terms 'photolithography' and 'microelectronics' are all products of the D.O.F.L.'s involvement with microcircuits at this period.[25]

Ultimately of vastly more importance than microcircuits were integrated circuits of the type developed by Kilby and Noyce in the late fifties. Though the general ethos of miniaturisation had been prevalent since at least the late forties, the actual concept of placing a complete circuit in one chip of semiconductor can be dated more precisely. In 1952, G. W. A. Dummer, from the Royal Radar Establishment at Malvern, delivered a speech in Washington in which he referred to the next logical step after the transistor being 'electronic equipment in a solid block with no connecting wires. The block may consist of layers of insulating, conducting, rectifying and amplifying materials, the electrical functions being connected directly by cutting out areas of the various layers'.[26] That is a very fair description of an integrated circuit and, for 1952, when industry was still piecing together transistors with tweezers under microscopes, quite remarkable.

By 1957, R.R.E. had engaged in a contract with the Plessey Company for the development of solid state circuits and a model of a solid state circuit in silicon was demonstrated at the International Symposium on Components held that year at Malvern.[27] American interest in this model was intense, much greater than British, and when the British Government was not forthcoming in funding the development of this integrated circuit, it was quietly shelved.[28] Yet, there was little separating the British model of 1957 from Kilby's working circuit, patented in 1959. Neither Kilby nor Noyce were aware of the work of Dummer or that of R.R.E. until the early sixties.[29] Failure to develop the R.R.E. model has been heralded as yet another British missed opportunity,[30] but there was a massive gap between the British model prototype – or even Kilby's integrated circuit for that matter – and the integrated circuits of the mid sixties. That gap was filled by the planar process, which allowed the integrated circuit commercial reality in a world in which anything less hardly counted.[31]

> Dummer was preaching the gospel of integrated circuitry as a gospel long before anybody, including Dummer, had the slightest idea how you could actually do this ... [Dummer] carried inspiration around on his back like pollen ... He never received the backing that his degree of inspiration would have justified. Alan Gibson

> Dummer's thinking was very valid and I think that, had we known about it, it might have accelerated the timing of things a little bit or a lot. Jack Kilby, Texas Instruments

Though all branches of the Military were interested in miniaturisation, the form that interest took varied among the Services. The Army and the Signal Corps supported micromodule schemes from 1958 based on the old Tinkertoy concept of stacked wafers. The Navy favoured the idea of thin films and funded an extensive research and development programme from 1958. The Air Force supported yet another idea; what is called molecular electronics, from 1959. The molecular electronics concept was radically different from any other attempt at miniaturisation. The idea was to build a circuit in the solid without reproducing individual component functions. The whole was to do more than the sum of the parts.[32] That the three Services should each choose to back entirely different approaches to miniaturisation was a product of their individual requirements and of the considerable controversy which surrounded the supposed prospects of each approach. It is also likely that a degree of inter-Service rivalry was not unimportant, though whether this resulted in healthy, pseudo-commercial competition or a waste of money, is a moot point.[33] What is less debatable about the part played by the Military in sponsoring miniaturisation work is its timing. Some observers see it as no coincidence that heavy military funding should have commenced so soon after the launching of Sputnik in 1957. Sputnik was as much a blow to American prestige as it was a triumph for Russian technology and especially Russian electronics miniaturisation.[34]

> I think all these ideas [for early integrated circuitry] essentially had to be simplified for the people who were funding the things and it's quite possible – from my own experience – that the Military didn't really understand what the details were. In fact, I'm quite sure they didn't ... All that you had to do was wave the Russian threat and you could get money. Douglas Warschauer

Molecular electronics demanded a technological leap beyond even integrated circuits, which were really just discrete components formed in the same piece of silicon with the twin problems of isolation and interconnection overcome. Molecular electronics was something else altogether, much more sophisticated and, to this day, barely possible. Even the name, with its incorrect implication that the behaviour of molecules rather than of electrons was to be controlled, met with criticism.[35] So too did the fact that the Air Force gave its entire $2 million contract to Westinghouse,[36] perhaps because of Westinghouse's interest in this area since 1957,[37] perhaps because no other company was interested in taking such a leap into the technological dark.[38] As Kilby has suggested, it was never very clear exactly what the Westinghouse programme entailed, except that it meant adopting an approach which totally disregarded two decades of experience with circuits and expecting to produce a revolutionary invention to order.[39]

> The Air Force rejected anything that had any connection with existing circuits ... They didn't want to get there circuit by circuit. They wanted these new breakthrough devices that would eliminate all that jazz.   Jack Kilby

In the early sixties it was not at all clear which of the several diverse approaches to miniaturisation would predominate, nor was there any consensus on how long a technique would take to become commercially important. Of the various approaches, the Army's micromodule programme seemed least promising, yet it was heavily funded as late as 1963 and was then seen as being an evolutionary technique which would be of service until integrated circuits became cheaper. At that time, it was still thought that integrated circuits would adopt a modular approach similar to the micromodule.[40] The way IBM chose to connect discrete components in its new 360 range of computers was, indeed, closely similar.[41] Few, apart from the Army and the Signal Corps, took the long-term prospects of micromodules seriously. Microcircuits, made of thin films deposited on ceramic substrate and incorporating semiconductor elements, seemed a much more worthwhile option in the early sixties. Indeed, these have been developed to the point where they are used in conjunction with integrated circuits to provide hybrid circuits for specialised applications.

Any detailed assessment of the future of miniaturisation as envisaged in the early sixties is rendered extremely difficult by widespread confusion over terminology.[42] Microcircuits, for example, could refer to anything from micromodules to integrated circuits,[43] and were sometimes represented as being the same thing as molecular electronics.[44] Perhaps the most common opinion at this time was that microcircuits would represent a transitional stage on the way to integrated circuits.[45] The future, it was held, would belong to solid circuits, a term which encompassed both integrated circuits and whatever might come of molecular electronics,[46] but nearly always the implication was that this domination would still be some years away.[47] Figure 8.3 gives an idea of the degree of miniaturisation that has taken place since 1940. It seemed natural to think in terms of evolution – from valve to transistor to micromodule to microcircuit to integrated circuit to molecular electronics.[48] The power of commercial forces and the importance of the production process were not to allow such logical development.

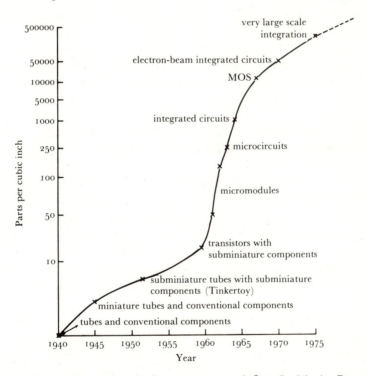

Fig. 8.3. Electronics miniaturisation, 1940–1975. (After G. W. A. Dummer, 'Integrated electronics', and S. H. Hollingdale and G. C. Toothill, *Electronic Computers*[83])

Planar techniques did not simply make feasible integrated circuitry, they shattered all possibility of electronic evolution.

> Certainly at the time [1961] I did not visualise anything comparable to a one-chip calculator or that level of complexity in the foreseeable future.   Jack Kilby, Texas Instruments

Table 8.1 shows the rapid increase in the number of integrated circuits produced and, more strikingly, the increase in terms of cir-

Table 8.1. *Shipments of discrete components and integrated circuits*

	Diodes and rectifiers (M)	Transistors (M)	Integrated circuits (M)	Discrete equivalents of integrated circuits (M)	Total discrete components or equivalents (M)	Percentage of components in integrated circuits
1963	627.1	302.9	4.5	108.0	1038.0	10.4
1964	762.1	398.0	13.8	331.2	1491.3	22.2
1965	1072.6	631.7	95.4	3052.8	4757.1	64.2
1966	1520.8	877.3	165.0	5280.0	7678.1	68.8
1967	1461.1	792.1	178.8	8582.4	10835.6	79.2
1968	1618.8	951.7	247.3	11840.4	14410.9	82.2
1969	2099.7	1249.1	423.6	23721.6	27074.4	87.6
1970	1866.0	976.9	490.2	27451.2	30294.1	90.6
1971	1473.1	880.7	635.2	40652.8	43006.6	94.5

Source: Jerome Kraus, 'An economic study of the US semiconductor industry',[75] pp. 80, 209.

cuit equivalents. Clearly, production of discrete components on this scale would have been rather difficult. Massive price reductions associated with increased production and learning economies had already assailed the industry in the early sixties and the intensification of this syndrome in the later sixties would have been intolerable.[49] Even had it been technically feasible to make and use the equivalent quantity of discrete components that were actually incorporated in integrated circuits by the seventies, past experience suggested that their production would not have been commercially feasible. But, in fact, in the same way that electronic systems of the forties were restricted in their sophistication and complexity by the limitations of the valve, so the systems of the sixties were beginning to be restricted by the limitations inherent with discrete components. These limitations showed up particularly clearly in military electronic systems where small size and high reliability were often of paramount importance. Hence, the Military was an important early market for integrated circuits and in the beginning virtually the only market. This is shown clearly in Table 8.2, together with the extremely high price early integrated circuits commanded.

Table 8.2. *Average uncorrected price of integrated circuits and proportion of production consumed by the Military*

	Average price ($)	Percentage consumed by the Military
1962	50.00	100
1963	31.60	94
1964	18.50	85
1965	8.33	72
1966	5.05	53
1967	3.32	43
1968	2.33	37
1969	1.67*	
1970	1.49*	
1971	1.27*	
1972	1.03*	

Source: John Tilton, *International Diffusion of Technology,*[64] p. 91.
*EIA, *Market Data Book* (1975)[84], p. 86.

This thing [a very early integrated circuit from Texas Instruments] only replaces two transistors and three resistors and costs $100. Aren't they crazy!

Attributed to a Philips director by P. W. Haayman, Philips.

By the early sixties, then, discrete components were approaching intrinsic levels of perfection in reliability, small size, performance and, above all, cost. Certainly a failure rate of, say, one per billion device hours could be improved, but not by much. As long as each element had to be made, tested, packed, shipped, unpacked, retested and interconnected with others, it would be the sheer individuality of components rather than technical or production limitations which would constrain improvement. The problem posed by the interconnection of components was particularly severe for, no matter how reliable the components, they were ultimately only as reliable as the joints connecting them and the generally manual methods used for wiring circuits. The more complex the system, the more interconnections were needed and the greater the chance of failure through this cause. Hence, the main obstacle to progress was a tyranny of numbers.[50] By 1962, some large computers contained as many as 200 000 separate components and future systems that would require 10 million or so were envisaged.[51] Not surprisingly, problems of reliability became serious with systems of this complexity. These could largely be overcome, as they were in the Minuteman project, by taking extreme care, by testing and checking, by keeping detailed records on each component and by duplicating vital systems, but such measures were not economic in the commercial world. It was said that if all military components received the cossetting given to those in Minuteman, the expense would have exceeded the gross national product.[52] Table 8.3 shows the annual number of computer systems delivered in the United States between 1955 and 1968 and gives a rough idea of the accelerating demand for the computer. What the figures do not show, though this is suggested by the rapid growth in the sales of components, is the increasing complexity of succeeding computers. Such complexity would not have been possible at an economic price without integrated circuits. Such circuits were not commercially available until 1962, but an early Texas Instruments exhibition piece of 1961 demonstrated just what such circuits could do. It was a computer which weighed 280 grammes, was 100 cc in volume and con-

Table 8.3. *Growth of computer production*

	Computer systems shipped	Cumulative number in use
1955	150	250
1956	600	800
1957	1000	1700
1958	1400	2950
1959	1700	4500
1960	2200	6500
1961	3400	9300
1962	4500	12600
1963	5600	16800
1964	7500	24000
1965	7400	31000
1966	10200	40600
1967	18700	57600
1968	14700	69400

Source: Jerome Kraus, 'An economic study of the U.S semiconductor industry',[75] pp. 80, 209.

tained 587 components. This replaced a predecessor which had 14 times as many parts, was 150 times larger and which weighed 48 times as much.[53]

One of the main worries of those critical of integrated circuitry was production yield. If yields were poor for discrete components, it was argued, then the failure rate would be magnified to immense proportions when attempts were made to fabricate more than one component in a single piece of silicon. If the yield for individual transistors was 50 per cent, then an integrated circuit on which there were four transistors would have a yield of $(0.5)^4 = 6$ per cent. But, if the yield of individual transistors were high, say 90 per cent, then the integrated circuit yield would be $(0.9)^4 = 66$ per cent – a significant improvement.[54] In fact, although high yield was as important as in the production of discrete components, it was nowhere near as critical as these figures suggest. Failure in production did not occur at random; often whole circuits and whole wafers would be ruined and could be discarded without further processing.[55] Had failure been totally random, the future of the integrated circuit would have been bleak indeed.

Integrated circuit yield remained a serious problem, though, throughout the first half of the sixties. Wafers used were generally larger than those used for discrete components and this increased risk of contamination, of cracks and holes interfering, of breakage in handling, as well as the difficulties of accurate mask alignment demanded by the planar technique. Although the manufacturing process was identical, the integrated circuit required more steps – something like twenty-two compared with perhaps fourteen to make a transistor. Thus, a 95 per cent yield for each step would produce transistors at 50 per cent yield, but integrated circuits at only 25 per cent yield. Until such yield problems were overcome, the industry could only compensate with high capacity and high prices for good circuits. It was obvious to some that prices would eventually decline and that huge potential production would encourage this, perhaps to the extent of the vicious price-cutting and over-production of discrete components between 1961 and 1963. Improved yields were bound to arrive. In 1963, for example, Motorola found that reducing wafer area to a quarter of its previous size in fact improved yields by more than a factor of four, simply because wafer defects were not randomly distributed.[56] This in itself was considerable motivation to make small circuits.

Another, and perhaps more general, concern was that the integrated circuit would bring serious change to the electronics industry itself.[57] Many circuit designers, it was imagined, would become redundant and the remaining designers would work within the component manufacturing companies rather than in the systems companies, those which had previously bought and assembled components. Thus, the new circuit designer would have to become much more knowledgeable about fields outside circuit design for he would need to consider not just what systems it was possible to put together, but what systems it would be possible and profitable to manufacture as integrated circuits.[58] It was also imagined that either those companies which had previously assembled components made by others – the systems houses – would start to make integrated circuits or that the component manufacturers would themselves become systems houses.[59] A particularly strong feeling was that the manufacture of integrated circuits would be on such a vast and expensive scale that only the largest companies would be able to afford to participate. Hence, the dozens of companies which had participated in the price-cutting of discrete components in the early sixties would be replaced by perhaps five huge companies supplying nearly all integrated circuits.[60] These things have happened to a degree, but certainly not to the extent anticipated. Circuit designers have had to become more closely associated with manufacturing, but there also grew up a new demand for designers of hybrid circuits assembled by companies which formed an additional layer between the component manufacturers and the systems houses.[61] Certainly some systems houses have begun manufacturing integrated circuits and some component manufacturers have also started making and selling systems, but neither has dominated and the situation remains fluid. It is also true that the cost of entry into the integrated circuit business has been high, but it does not seem to have been absolutely prohibitive. As Table 8.4 shows, as many smaller firms remained active in the semiconductor industry as at almost any other time, though by no means all manufactured integrated circuits.

The guy that sells components doesn't make any money. You gotta wire them into something before you can get to make some money out of it . . . The component seller gets beat to death on price . . . The curve of prices has far outstripped the curve of costs over the past twenty years. We have made up for a lot of it by making things more

Table 8.4. *Concentration of semiconductor shipments by companies (percentage)*

Total shipments by	1958	1962	1965	1972
4 largest companies	51	42	50	50
8 largest companies	71	65	77	66
20 largest companies	97	89	90	81
50 largest companies	100	95	96	96
All companies	100	100	100	100

Source: William Finan, *International Transfer of Semiconductor Technology,*[82] p. 7.

> complex, by getting them onto a smaller piece of real estate, that type of thing.   William Winter, Westinghouse

> There has had to be a measure of compromise. Obviously in the early days the systems people, experienced in ordering discrete transistors and having to optimise the way they used them in circuits, got used to having fairly large degrees of freedom, I think. Now they are in a situation where they have a package and they can't have the same degree of freedom. They came to appreciate that that was a very small price to pay for having these large chunks of highly complex circuits available so cheaply and with such reliability.   John Cave, Plessey

Observers of the early sixties generally underestimated the size of the total electronics market of the early seventies,[62] but overestimated the future importance of the Military as a market for electronics products. An industry survey of 1962 foresaw that the Military would buy over 57 per cent of all electronics sold and nearly half the integrated circuits produced by 1970.[63] In fact, the Military bought about 40 per cent of all electronics by that date and about a third of the integrated circuits manufactured.[64] A more elaborate and more expensive survey in 1963 was equally incorrect in predicting that at least half the integrated circuits produced in the early seventies would go to the Military. It was, however, much more accurate in its estimation of the size of the integrated circuit market, and in its emphasis on the future importance of cost. Price reductions, more drastic even than those for transistors, were predicted and these certainly came to pass.[65]

There was also a technological battle within the field of integrated circuits throughout the sixties. The original integrated circuits were, like the transistors of the fifties, bipolar, with electron action taking place within the body of the semiconductor. Yet it was possible to achieve transistor action at the semiconductor surface in the way sought by Shockley and his many predecessors – the field effect. RCA had had some success in preventing contamination, the main production difficulty with this sort of device, and Fairchild had also shown interest in 1962, but both companies lost impetus because of their primary interest in bipolar technology.[66] This allowed two small firms, General Microelectronics and General Instrument, to enter the field.[67] Both met with serious production difficulties in the mid sixties, but by 1970, the MOS (metal oxide semiconductor) integrated circuit had become a viable product. It was much slower than bipolar integrated circuitry, but because it required much less power, more integration was possible, and because it required fewer processing procedures, it was cheaper to make. Therefore, for calculators, computer terminals and small electronic machines, MOS was ideal and became the technology by which both large-scale integration and large-scale price-cutting were effected.[68]

MOS permitted more and more circuits to be put onto a piece of silicon using processes which were based on planar methods of the early sixties. Since 1960, the number of components it is possible to put on individual chips has about doubled every year.[69] The average number of active components built into chips is shown in Table 8.5 and falls short of the possible, but there has clearly been a steady increase in the degree of integration obtained. The degree of integration is measured in bits or gates (a circuit which fills a logic function) per chip. While the exact meaning of terms is as imprecise here as it is in most other areas of microelectronics, a majority would probably agree that Medium-Scale Integration (MSI) refers to a chip containing between a dozen and one hundred gates or bits per chip

Table 8.5. *Average number of components per chip by year*

1963	24	1968	48
1964	24	1969	56
1965	32	1970	56
1966	32	1971	64
1967	48		

Source: Jerome Kraus, 'An economic study of the U.S. semiconductor industry',[75] p. 208.

and Large-Scale Integration (LSI) to chips containing more than one hundred,[70] though some put the threshold at fifty bits per chip.[71] Large-scale integration is possible only with MOS and its growing use is attributable to the re-emergence of the tyranny of numbers problem which plagued the discrete components industry of the late fifties. Electronic equipment gradually became more complex until it was using as many simple integrated circuits as it had been discrete components. Consequently, the same problems of interconnections, testing and packaging arose and were solved by placing more and more components on one slice of silicon.

The MOS integrated circuit permits intense integration – by 1971 it was possible to put 6000 component equivalents on a chip about 6 mm square[72] – but this obviously means that space on the chip is at a premium. There is thus a need to produce circuits which use the maximum number of small active components, such as transistors and diodes, but the very minimum of the much larger passive components, such as resistors and capacitors. This is the very opposite of requirements for circuits using discrete components, for discrete passive components are easy to make and cheap while active components are more complicated and expensive. It is also the reverse of thin film microcircuit requirements, for with these the formation of passive components was even easier and of active even more difficult. Hence, the sort of circuits most amenable to integration were digital circuits using few passive components but many active ones to perform a logical 'yes–no' function. This is precisely the sort of circuitry demanded by computers and demanded in the vast quantities in which the larger companies were anxious to produce. Hence, mass produced LSI circuits are digital while the more complex linear circuitry, requiring more passive components, is produced using discrete components or using relatively simple integrated circuits (containing active components) attached to films (on which it is easy to produce passive components) to form a hybrid circuit.[73] In 1968, the value of the discrete transistor market, though declining, still exceeded total integrated circuit sales. Within the integrated circuit market, digital circuits outsold linear by $304 million to $60 million,[74] though MOS then had only a tiny portion of this market.[75]

There is some feeling that the prospects of LSI are limited. When thousands of circuits are packed onto a chip, there is a tendency for chip size to increase and for yield problems to increase accordingly.

Attempts to improve yields by isolating faulty circuits using discretionary wiring techniques, whereby a computer selected the good circuits on a wafer and connected only these, have not proved successful.[76] But a more fundamental problem is that the more integrated the chip, the more specialised its function. It is unlikely to find a mass market and such a market is absolutely essential for the batch production methods and long runs which seem to be the only economic way of manufacturing such integrated circuits. It is suggested that the greater flexibility offered by MSI and the hybrid approach offer much more scope for the future,[77] that LSI is most useful as a proving ground for MSI[78] and that it is no more than a synonymous term for 'custom design'.[79] Such estimates of the situation may be quite correct, but if they are not, it will be because they have ignored the importance of price. MOS is the cheapest form of integrated circuitry and LSI is the most exhaustive exploitation of that form. The past significance of low price in this industry needs no further emphasis here. Certainly LSI must involve electronics firms more closely in final systems, as happened as early as 1968, for example, when General Instrument combined with Hammond to produce 450 bit integrated circuits for organs.[80]

In the same way that a multiplicity of transistor types was produced for a huge variety of applications in the late fifties, so integrated circuits offer a wide variety of circuit functions for even more applications. Where price is low, as it is with MOS and LSI, there will be an enormous increase in the number of applications as it becomes economically feasible to do more and more things electronically. Instead of integrated circuits tailored to fit possible applications, the very low price of some integrated circuits may well attract new applications suited to what the circuits can do.[81]

> Any product that uses springs, levers, stepping motors or gears is performing logic and that product should be built of semiconductors.
> Floyd Kvamme, National Semiconductor

> More than any other single area of technology that man has ever conceived and exploited, electronics in general and, more specifically, integrated solid state electronics, will have more and wider impact on more individuals in the world than any other technology that has been exploited.
> Dean Toombs

CHAPTER 9

The American semiconductor industry

The electronics industry is one of the major industries of the world and much of it is now a semiconductor industry. Other huge industries, such as aerospace or computers, are now largely or utterly dependent on semiconductors. The American industry dominates totally the world scene in the production of these semiconductors. Table 9.1, comparing U.S. semiconductor production with total

Table 9.1. *World semiconductor consumption and U.S. shipments (uncorrected values)*

	World consumption ($M)	U.S. shipments ($M)	U.S. production as percentage of world consumption
1969	2217	1218	54.9
1970	2324	1119	48.1
1971	2211	1066	48.2
1972	2750	1350	49.1
1973	4080	2200	53.9
1974	4500	2550	56.7
1975	3875	1900	49.0
1976	4846	2730	56.3

Source: Coleman & Co. estimates, 1974; WEMA Semiconductor Forecast, 1974.

world consumption, illustrates this vividly, though it should be remembered that much of this consumption takes place in the United States.

Why is it that the American semiconductor industry has become so powerful within the United States and internationally? Certainly the advantage of a huge domestic market is a primary factor contributing to the success story, but other factors must be sought to explain the international domination of the Americans. Has this situation arisen simply because the transistor was invented in the United States, so giving that country's industry a head start which it has never lost? Has extensive American military support of the industry been a primary factor in its success? Perhaps the environment in which the American semiconductor industry has developed has played a crucial part in its success. Indeed, perhaps the industry has been so successful that it has created its own environment especially suited to its needs, thus ensuring that success breeds success.

We have seen how the semiconductor industry of the early and mid fifties was dominated by a handful of large, established electronics firms and how their position was rapidly eroded by new companies entering the semiconductor business. By the end of the decade, there were dozens of firms in that business[1] and the market leaders were not the established companies, but new ones.[2] The older firms remained the innovation leaders, but in quantity of innovation rather than in its exploitation.[3] By the late fifties, the discrete semiconductor market had itself become dominated by two factors; high production capacity and low price, a pattern that was to repeat itself in the sixties with integrated circuits. Between 1961 and 1965 average integrated circuit price declined 80 per cent and the number of companies producing them rose from two to over thirty.[4] Some firms left the industry and others joined so that by 1972 there were about 120 semiconductor companies, approximately the same number as a decade before.[5] Table 9.2 gives a division of this number into large and small firms, the criterion being a fairly arbitrary one of semiconductor sales for 1972 exceeding $20 million or total corporate sales, including semiconductors, of over $100 million. Captive firms are those which are part of corporations with other interests. Clearly, most small firms are independent of such

Table 9.2. *Size and type of U.S. semiconductor firms in 1972*

	Large	Small	Total
Captive	24	7	31
Non-captive	25	43	68
Unknown	–	21	21
Total	49	71	120

Source: William Finan, *International Transfer of Semiconductor Technology*,[4] p. 6.

corporations. The small firms have confounded those observers who prophesied that they would be displaced by larger companies as the industry matured. In 1958, the largest twenty semiconductor companies had captured 97 per cent of the market: in 1972, the top twenty had only 81 per cent.[6] Certainly the very largest companies exert enormous influence on the market – the biggest four control about half of it – and this has nearly always been the situation. What has changed radically is the identity of the top companies. In 1971 the top five companies were IBM, Texas Instruments, Motorola, Western Electric and Fairchild,[7] not one of which had been a leader in the commercial semiconductor market in the mid fifties.

The location of the semiconductor industry has also changed significantly. In the early fifties, it was an East Coast industry because that was where existing large electronics firms were located. Later that decade, as many small firms burst into the market, it remained on that Coast but became nucleated in those areas which best suited the demands of the new industry. Long Island and the Boston region, particularly the area around Route 128, proved most attractive to new companies at this time and many semiconductor firms are still to be found there.[8] Several reasons are commonly given in explanation. These areas offered a wide range of ancillary industries, the many skills and products of which were seen as being useful to the infant semiconductor industry. The cities of Boston and New York were, in the broadest sense of the term, cultural centres near which there was some incentive to live and work. They were also the homes of major universities and to an industry exploiting the frontiers of science, this seemed important. Many of the founders of new companies were originally graduates of such institutions as M.I.T. and Harvard, other staff were readily available from the same sources and frequent consultations with university experts were possible. But of even greater importance than any of these was the availability of money. Any new commercial venture needs financial support until it is strong enough to look after itself. There is always a chance of failure and the risks are particularly high for a company based on a revolutionary and unproven technology. Such financial backing is known, most appropriately, as 'risk capital' and its availability in such areas as Boston was probably the crucial factor in the early concentration of semiconductor firms. A survey of founders of new science-based firms in Philadelphia and Boston revealed that the proximity of science-oriented universities and the availability of risk capital made Boston a very much more attractive site than Philadelphia.[9]

In the late fifties, when low prices made the problem of yield particularly acute, there was some interest shown by semiconductor companies in moving to areas where geographic conditions seemed most suited to semiconductor processing. Honeywell, for example, had planned to build a semiconductor plant in Colorado, where there would be clean water, fresh air and low humidity. Such factors seemed essential when much of semiconductor processing was a black magic art, but they could also be created artificially in purpose-built semiconductor factories, equipped with de-ionisers for water, dry boxes with filtered air for assembly and low humidity rooms for photoresist work. Such equipment would also have been necessary in Colorado.[10]

The location requirements of the semiconductor industry were not to be entirely satisfied by the criteria most other industries judged important. Much of the semiconductor industry is still concentrated in specific areas, but the East Coast is no longer the centre of the industry. Instead, the West Coast has attracted semiconductor companies and the hub of the industry is now California, and in particular San Francisco. Just to the south of that city is the Santa Clara Valley, known as 'Silicon Valley' because there were, by 1969, no less than twenty-five semiconductor firms located there within a few miles of each other.[11] It is likely that there are now even more. The reasons for this flourishing of semiconductor companies in the San Francisco Bay region are well worth some detailed exploration and are central to any discussion on the success of the American semiconductor industry.

The first semiconductor company in the San Francisco area was founded by William Shockley at Palo Alto. Shockley had been tempted by the opportunities that seemed to exist in the commercial world for his mastery of semiconductor science. A brief affair in 1954 with Raytheon, one of the old established companies on the East Coast, from whom Shockley wanted $1 million over three years as a consultant, came to nothing[12] and Shockley determined to set up on his own. Shockley Semiconductor Laboratories was established in 1955 as a subsidiary of Beckman Instruments and located at Palo Alto because that was Shockley's home town.[13] The company has had a confusing history, but that is typical of the industry. In 1958, it changed from a research unit to a manufacturing company and became Shockley Transistor, producing, in a quasi-university atmosphere, sophisticated diodes tested to well beyond military specifications.[14] In 1959, the firm was bought by Clevite, and then by ITT in 1965. By 1969, ITT had closed the Palo Alto factory.[15] Shockley's commercial career would be a matter of little significance had his reputation not attracted several of the relatively few very able semiconductor experts in the fifties to work for him. In 1957, eight of these men – all originally from the older electronics firms in the East, chose to leave in order to found their own company.[16] That company was Fairchild Semiconductor.

The influence of Fairchild is felt throughout the semiconductor industry, but nowhere more than in Silicon Valley, where most top semiconductor men have at one time worked for Fairchild.[17] At a conference held in Sunnyvale, California in 1969, of 400 semiconductor men present, less than two dozen had never worked for Fairchild.[18] Some forty-one companies have now been formed by former Fairchild employees[19] and, not surprisingly, many of these are in the Santa Clara Valley. There can be no doubt that the location of Fairchild explains much of the present concentration of the industry in California, but that is only part of the explanation. It could be argued in the same way that the California industry exists because Palo Alto was Shockley's home town. That is true, but hardly satisfactory.

The San Francisco Bay area has other attractions besides Fairchild. Not unimportant is that it is an extremely pleasant place to live and work. The general standard of living is high and the climate is ideal. Semiconductor buyers from other states are reputed to prefer doing business with California firms rather than those in, say, Boston when regular business trips are required.[20] The proximity of a cultural centre in San Francisco also seems to be important, though how critical such factors are in determining the location of industry is difficult to quantify. Certainly such personal factors should not be overlooked, particularly in this industry where the priorities of individuals have regularly been of fundamental importance. When Sprague sought to enter the integrated circuit business in 1968, one main obstacle was estimated to be the difficulty it would have in attracting top people to work in Massachusetts rather than California.[21] It is said that one firm eventually built a semiconductor plant in Florida because one of its directors happened to have a home there,[22] and that the site of another in Florida is not totally unconnected with the excellence of the local fishing.[23]

There are other reasons for the siting of so much semiconductor activity in the San Francisco area. The population of California has grown enormously in the last two decades, particularly in the conurbations, and much of this new population is young, well educated and adaptable. Many other high-technology industries, such as the aerospace industry, are now established in California. There is a reservoir of personnel to fill the demands of semiconductor firms for technicians and laboratory workers. While large numbers of unskilled assembly workers might be readily obtainable anywhere, a reserve of trained and experienced technicians is a rare commodity and one which seems to influence the semiconductor industry's preference for California.[24] So keen are even assembly workers to join the ranks of the technically qualified that at least one Santa Clara college now offers courses in semiconductor production technology for just such people.[25] There is even a school of opinion in the industry which holds that California workers are particularly suited to a young, aggressive and highly competitive industry, as they themselves are often the sort of people who have gone to some lengths in order to improve their lot in life and are likely to possess identical characteristics.[26]

A much more tangible reason for the growth and continuance, if not the emergence, of the San Francisco semiconductor industry is the presence of a large number of companies which, though not themselves semiconductor manufacturers, make the mass of ancillary equipment those manufacturers require.[27] A small semiconductor firm on Long Island was recently making its own testing jigs and preparing its own silicon powder because available engineering and chemical companies did not appreciate the very specific needs of the semiconductor industry.[28] Such a situation would have been much less likely in the San Francisco area, where ancillary industries are very much geared to meet the requirements of the semiconductor industry.

In theory, the proximity of the University of California at Berkeley and Stanford University should provide inducement for semiconductor companies to be located in the vicinity. Both schools have a high academic reputation and are considered leaders in the field of solid state electronics.[29] The theory is arguable, and not a few men in the semiconductor industry see the academic reputation of Berkeley and Stanford as a product of industrial progress and not vice versa. The

University was imagined to have followed where industry had led and there was sometimes little contact between the two.[30] Some men in the industry saw University solid state work as being almost parasitical in that courses were often taught by part-time professors from local industry.[31] This in itself is an attraction for those in the industry who wish to have both an industrial and an academic career and may indirectly increase the desirability of the area for semiconductor companies.[32]

The advantages one semiconductor company gains from the proximity of others are considerable. In such a highly competitive industry, it is vital to keep abreast of latest technological and commercial developments. While it would be going too far to say that knowledge is pooled among the San Francisco semiconductor community, it is quite reasonable to imagine a very rapid dissemination of knowledge. In the semiconductor industry, expertise is transitory and extremely valuable. The concentration of semiconductor companies in Silicon Valley provides excellent conditions for the transfer of knowledge. The industry is founded on expertise and survives by replenishing this expertise with the help of individuals possessing the latest, appropriate knowledge. Whether this is achieved by an informal chat on the golf course between employees of different firms, or by the hiring of an expert from another semiconductor company, the transfer of knowledge must always be more easily accomplished when firms are close together as they are in the Santa Clara Valley. Friendships are more readily made among people in the same business in regular contact with each other and a move to a new job is often much more attractive if it does not mean moving house, changing the children's schools and finding new friends. So important is contact between individual experts seen to be in this industry that the Waggonwheel Bar at the corner of Wissman Road and Middlefield Avenue in Silicon Valley has been spoken of – not completely in jest – as the fountainhead of the semiconductor industry. It is now probably past its heyday, but situated only a block from Intel, Raytheon and Fairchild and not much further from many other companies, it was a place where semiconductor men drank, exchanged information and hired employees.[33]

It is also of no small relevance to the health of the semiconductor industry in Silicon Valley that the area is one in which risk capital to finance new enterprise is more readily available than in other areas. Though much initial funding of semiconductor companies comes from private sources,[34] a receptive attitude on the part of banks and finance houses towards new high-technology industry has done much to encourage the proliferation of semiconductor companies in the area. A new breed of men, expert in both financial affairs and electronics, has arisen to assess the potential of new companies seeking to exploit new ideas. No matter how good the technology or the personnel, the new company is likely to experience severe difficulties getting started without capital and this can be hard to attract when both the company's founders and its technology are unproven. In San Francisco, financial institutions have come to understand the semiconductor industry and their support has been fundamental to its growth.

Certainly semiconductor firms are located in other parts of the United States. Many remain in the original areas of concentration in the North East and still value the attractions which brought them there, as well as links forged over many years with local suppliers and customers. The largest semiconductor company of all to sell on the open market, Texas Instruments, is sited in splendid isolation in Dallas where, some argue, its leadership in the industry renders its distance from the pack of little consequence.[35] It may be that Dallas is conducive to Texas Instruments rather than to the semiconductor industry in general. Few spin-off companies have emerged from Texas Instruments and of these few, at least one, Siliconix, has fled to the more amenable surroundings of San Francisco.[36] Various other companies are located far from the main stream. Motorola is in Phoenix and probably big enough to regard constant, intimate contact with competitors as being less than vital. Like Texas Instruments, Motorola has had few spin-off companies established in its neighbourhood. Semiconductor men working in Phoenix view its rather warm climate as an attraction and seem to value links with the University of Arizona.[37] Other semiconductor men working for firms on the geographical periphery of the industry do, however, regard their location as a disadvantage. Some of those working for Florida companies imagine that their distance from most other firms presents an obstacle to the flow of information.[38]

In considering factors responsible for the location of semiconductor companies, particularly in California, it has been necessary to emphasise the importance of conditions which semiconductor experts find attractive. The industry is one of people rather than of companies in that the prosperity of these companies is, perhaps more than in any other industry, dependent on the abilities of in-dividual experts.[39] It is, therefore, not unreasonable to stress the im-portance of cultural centres and a warm sun. From its very incep-tion, expertise in the semiconductor industry has been the property of individuals rather than of companies. In its early days, the skills required by the industry were acquired by hiring individuals, each of whom could cope with parts of the pioneer technology. Future progress was dependent on what those individuals, singly or in groups, made of the new technology. As we have seen, those who were most successful often found the conditions within the es-tablished electronics firms most frustrating and the rewards offered outside these firms most attractive. Consequently, many left to join other companies or to set up their own, often in collaboration with old colleagues. This is a fairly understandable phenomenon in the early days of a new industry, particularly a high technology in-dustry,[40] and the normal pattern would be for the value of the in-dividual to diminish as some products, processes or even companies prove their superiority over others and become established. But this has not yet happened in the semiconductor industry. The rate of change has maintained its pace and this has maintained the value of the individual.

Change in the semiconductor industry is very much a function of easy and regular contact with all others engaged in semiconductor work. If a company is to do more than tread water, it must keep abreast of latest developments and put itself in a position in which it can take advantage of them. As in other industries, this can be done by one firm granting a licence to another for the use of its patents. Such licensing is particularly prevalent in the semiconductor in-dustry because of frequent customer demand for a second or even third source of a product, but it seems that it is not just money which changes hands. The fee is as likely to be information and the main value of patents in the industry is in the bargaining for and exchange of information. Patent rights are generally ignored until the sales volume of the offending firm begins to attract the attention of the patent owner.[41]

While those semiconductor experts working for government research laboratories or in universities usually cite the more formal avenues of communication, such as published papers, reports or conferences, as being of most importance to them, those in industry seem almost unanimous in recognising the primary importance of less formal channels. Published papers, for example, are often poorly regarded by those in industry because it is assumed that they will say nothing of commercial value.[42] Employees of rival companies will consult by telephone across the country. Company secrets are unlikely to be disclosed, but much useful knowledge in the industry arises from personal experience, and this is rarely reckoned the exclusive property of any institution. It pays to help someone who will almost certainly be in a position one day to help you. Such contact makes a nonsense of the strict security with which most semiconductor companies surround themselves. So too does the personnel mobility within the industry. At one time it was fairly common to demand that employees sign a promise not to divulge information for a stated period after leaving a company. That is now generally recognised as totally impractical.[43] Where large numbers of employees have departed together from one company for another, as when Lester Hogan left Motorola for Fairchild with the entire senior management of Motorola's semiconductor division, legal action has been taken, but this has nearly always been protracted and inconclusive.[44] When at Motorola, Hogan himself had successfully attracted eighteen semiconductor engineers from General Electric.[45] In 1969, IBM succeeded in obtaining an injunction against the Cogar Corporation to stop it recruiting IBM personnel, but by that time 70 of Cogar's 150 employees had already been poached.[46] When Jean Hoerni, ex-Shockley Transistor, ex-Fairchild, ex-Amelco Semiconductor, ex-Union Carbide, hired half a dozen or so experts from National Semiconductor to join his new company, Intersil, National secured a consent decree to prevent future encroachment.[47]

> Most of the intelligence interplay that goes on in the industry is through guys quitting and whoring themselves to competitors . . . If I developed something that I thought was key to my business growth, the last thing I would do is give a paper on it.
>
> U.S. semiconductor executive

> I have people call me quite frequently and say, 'Hey, did you ever run into this one?', and you say, 'Yeh, seven or eight years ago. Why don't you try this, that or the other thing?' We all get calls like that.
>
> William Winter

> I found that if a guy tells you something, it's probably straight. It's just whether or not he'll tell you.   U.S. semiconductor executive

The older established electronics firms and Bell Laboratories supplied many of the experts who sought to use their talents elsewhere in the fifties. They continue to disgorge such men, but the supply has been considerably augmented by the exodus of talent from the newer semiconductor firms. Fairchild, the creation of dissidents from one new company and the supplier of veritable legions of recruits to other semiconductor firms, is obviously the supreme example. But mobility in the industry is so widespread and accepted that it would be misleading to imagine that Fairchild is exceptional in anything but degree. Very few semiconductor experts have spent their careers with a single firm and very many have worked for three or four within a decade. At one extreme may be the man who has worked for many years for one of the older companies in the East and who values the security.[48] At the other are the 'job-hoppers' of the industry, men who work for one firm long enough to learn a technique and then move to another to sell their knowledge and to acquire more. Such 'turn-key' knowledge is naturally self-propelled, for firms employing its holders understandably have little use for such men once they have disclosed their secrets. It is possible for someone who knows really very little about semiconductors to do well by selling himself from firm to firm,[49] but some companies place a low value on such knowledge[50] and the price of such an expert can be very low indeed if he has been out of work for some time.[51] Most semiconductor experts lie between these two extremes and it is well worthwhile examining the reasons for their extraordinary mobility and for the strength of personal motivation in the industry as a whole. Both seem to have been essential to the rapid progress that has been made.

There is absolutely no doubt that one main motivation is money. A man will frequently take his expertise where it will be worth most to him. While a large company may be in a position to make the most of an individual's technique, it may not be able, or choose, to offer that individual the financial rewards available elsewhere. Hence, the individual may move to a more generous firm or, if he has the confidence and capital, start his own company based on his bright ideas. Promotion above the salary ceilings that often exist in large companies may be slower and more difficult than starting a new company or joining an existing small firm in which salary ceilings (or floors for that matter) do not exist.[52] Some individuals in the industry earn a great deal of money and a few have made fortunes. A new small firm which has gone some way towards developing its own promising technology may well be bought by a larger concern, making rich men of the small company's founders. When Robert Noyce sold his share of Fairchild to the parent company in 1959, he is said to have made $600 000 out of the deal and when Phil Ferguson sold his part of General Microelectronics to Ford in 1966, he apparently profited by no less than $300 000.[53]

> The top few people in the company have the opportunity of making a great deal of money if they are successful, and I think this provides the motivation to work night and day and to really just out-perform the competition ... I wouldn't have left Fairchild except for the prospect of making some money. I just wouldn't have done it because that was interesting too. I had a good job and a good future there.   U.S. electronics consultant

Semiconductor firms attract and reward valuable employees not so much by offering high salaries as by giving stock options. A man will be offered so many shares in the firm at a favourable price, shares which are redeemable at a future date. He may even be lent the money to purchase the shares. The firm gets the individual and his interest in the prosperity of the company: the individual can sometimes make a great deal of money. When Lester Hogan moved from Motorola to join Fairchild in 1968, he was offered a salary of $120 000, an interest-free loan of $5.4 million to exercise an option on 90 000 shares at $60 each, and a further allotment of 10 000 shares at $10 each. By late 1968, the paper profit on the shares was estimated at $2.5 million.[54] This transaction so impressed the industry that subsequent major transfer deals were sometimes measured in units of Hogan. A man would be said to have changed companies for a half or a quarter Hogan.[55] Naturally, not all deals brought such prosperity, stock options being worth no more than the company which issued them. In a large company, the distant relationship between personal effort and stock value often makes stock options less attractive than in a small company.[56] When companies have offered no stock options at all, they have sometimes seen their employees desert in consequence. Groups certainly left both Fairchild and Transitron for this very reason in the days before those companies changed their attitudes towards stock options.[57]

> The semiconductor world is populated by almost rich professional managers who took jobs with a nice salary expecting to become millionaires with the stock option they got and every one of us has

somewhere in a dresser drawer stock options worth garbage and not that much money in the bank.　　　　　U.S. semiconductor executive

It would be naive to see money as the only, or even the main, factor motivating men in the semiconductor industry and encouraging their mobility within it. Many leaders in the industry are scientists and technologists rather than industrialists or businessmen – a situation which will be seen later to have brought its own problems – and are responsible for much of the furious pace of its technological innovation. Not surprisingly, many become fascinated by the challenge offered to their scientific and technological skills. In the same way that a university scientist might work day and night in his laboratory to solve an important, intriguing problem, so some men in the semiconductor industry become engrossed in their work. When their employer refuses to match their interest, or when an exciting discovery is made, there is a temptation for such men to go elsewhere that has little to do with money.

> Most good people like to see their ideas go someplace besides into a memo. They like to see it come out as a product . . . . You would be surprised at the number of professional people who do not work for the almighty dollar, I mean that really work for self-satisfaction.
> Kenneth Manchester

> I was absolutely amazed at the number of engineers who were interviewed for jobs on the development side who wanted to meet the marketing guy. They wanted to have some assurance that their zippy ideas that were running around in their head were going to in fact get marketing push. Now where do you find that elsewhere?
> Floyd Kvamme

> I've never been bored in this industry – never . . . This is a state-of-the-art type of industry and the state-of-the-art is what kind of a recipe you can put together to do it right. That's why you never get bored.
> Kenneth Manchester

The marked progress of the industry encourages a feeling among individuals that they are each part of a winning team. The numerous and often spectacular consequences of semiconductor technology are seen as justification for considerable effort and some pride is felt in having played even a small part in producing major new technology. Because their product is important, those who have created the product feel important, and probably with some justification.

I haven't the least idea whether I could make a great income doing something else, but this is a lot of fun and there's a lot of satisfaction in the sense that you have changed the future of the world . . . I had the people who did the solar cells for Telstar.   James M. Early

We knew what we were using our devices for . . . It was quite exciting when the picture was coming back from the Moon, to say to my friends and my kids 'We helped bring back that picture'. I found that very motivational and that's been the story of my career. I've always been tied to something and known where the devices were going to be used and I've found that rather exciting.   Gene Strull

Semiconductors are fun . . . The fun part of it is that you are really using your brain . . . The industry is young enough so that all solutions aren't tried and true . . . The personal achievement I feel in the whole thing is that in some ways I have contributed to a technology that is really serving man. I have had a little piece of the action.   William Winter

That's one of the exciting things about the industry; a day never seems to go by where you don't hear of or read of a new application where semiconductors are used.   John Rogers

The barriers which sometimes separate the businessman from the scientist/technologist in modern industry are rarely evident in the semiconductor industry and it is not uncommon to find semiconductor experts who are involved in both spheres. This is an unusual combination and one which many in the industry find attractive. There are drawbacks, but the advantage of increasing an individual's involvement with the whole of a company's activity rather than just a part is important. Men feel that more use is being made of their abilities and are less likely to see themselves as small cogs in large machines.

Many of us, at least in this industry, start out as technologists, or engineers and then we become more and more interested in the business or economic aspects of any particular corporation. Semiconductors allow you to do both.   Mike Callahan

Because of its rate of change and its competitiveness, the industry is exciting. It offers long hours, hard work and a constant challenge, but this is manna to some. The industry attracts the well qualified – of the eight who formed Fairchild, six had Ph.Ds.[58] – and can bestow its rewards on the young. Robert Noyce was only 32 when he became head of Fairchild Semiconductor.[59] Such men inspire confidence and attract other good people to them.[60] In semiconductors, success breeds success.

> I like to participate in the game in order to see if I can win.
>
> Robert Noyce

As in other industries, there are generally fewer places at the top than people with ambitions to get there. The rapid rate of technological change in the semiconductor industry and the fact that much of this has been the product of the inspiration of individuals have opened an alternative route to the top. Semiconductor experts have frequently founded their own companies to benefit from the pleasure they feel in being their own boss.[61] It is likely that the industry has also benefited, for small firms started in just this way have a distinguished record. Some of this distinction may well be the product of the long hours of endeavour people often feel worthwhile when they are working for themselves.

Mobility within the semiconductor industry has been aided by the very tolerant attitude of most firms towards movement of personnel. Bell Laboratories, the source of so much expertise 'on the hoof', established a precedent in its lenient attitude to the departure of those who wished to work in the commercial semiconductor market despite the fact that they often carried Bell's technology with them. The primary explanation seems to be the value Bell places on the strong personal links Bell scientists now have with those working throughout the industry as a consequence of this outward flow.[62] If mobility is accepted at the company level, it has almost come to be expected at the level of the individual. A man who has not changed company is anxious to explain why: a man who has, perhaps several times, feels no need to justify his actions. Mobility has become the norm.

> We all know each other. It's an industry where everybody knows everybody because at one time or another everyone worked together.
>
> Floyd Kvamme

> You better watch out who you talk to and what you say to every person because he may be your boss some day.   Tom Mitchell

Small companies seem vital to the organisation of the American semiconductor industry. They may innovate themselves, and so lead the way to new products or processes, or they may, by their very existence, pose a threat which forces larger companies to innovate. They supply products for which there is only a small, specialised demand which does not attract the larger firms and they are able to provide the second source of supply which is a normal demand of those buying semiconductors.[63]

> The parts of the business where big companies have been successful
> in the past have been those areas where the technology was not
> moving rapidly . . . They do all right if the technology is mature and
> not changing, but it's the small companies who have been the ones
> who did all the innovation.   U.S. electronics consultant

> As long as there are possibilities for major new breakthroughs, there
> is going to be a role for small, new firms, I would argue, coming in
> and taking advantage of that because the big, established firms are
> going to want to concentrate more on the proven technology where
> their scale has some sort of an advantage.   John Tilton

> No matter what happens to technology, I feel really strongly that
> there will always be room for the little guy, making the little quantity
> of the specialised part for somebody.   Richard Gerdes

> Most companies have finally learned that you cannot successfully be a
> large manufacturer as well as an innovator.
>   U.S. semiconductor executive

It seems reasonable to see small firms in the semiconductor industry as an antidote to some of the problems an industry composed
purely or mainly of large firms would face. Many semiconductor
men complain that they cannot work satisfactorily in the atmosphere of a big company, particularly one in which decisions are
taken by distant managers of diversified corporations. When Hogan
took over Fairchild Semiconductor, he insisted on transferring the
headquarters of the whole Fairchild Corporation from Long Island to
Silicon Valley.[64] When such large corporations have tried to control
semiconductor divisions as though they were much the same as any
other division making any other sort of product, they have courted
disaster. This was true in the early days of the semiconductor industry when the established electronics firms failed to match the
pace and enterprise of the newer firms and it has been equally true
ever since. One manager of Union Carbide's semiconductor
operations has graphically outlined the difficulties. Union Carbide
had a huge income of $2000 million, but its interests were in
chemicals and minerals rather than its $10 million risk venture in
semiconductors. The company was capital intensive and could not
adjust to job-hopping, high salaries for valuable individuals and
what it called the 'California style' of business. The manager and
others simply left.[65] Several of the oldest and largest semiconductor
corporations, including Sylvania, Philco-Ford, Sperry, Westing-

house, Bendix and General Electric have recently deemed it prudent to withdraw altogether from the commercial semiconductor market, although some still manufacture semiconductor devices for their own use.[66]

> I had a tremendous rapid change from being the president of a little company that was totally entrepreneurial to selling that to and becoming part of a large organisation. You really can see why these companies just cannot be successful because they insist that the semiconductor operation operate in exactly the same manner as they run a steel mill ... Large companies' control mechanisms don't let them do something until it's obvious and by that time it's too late.
>
> U.S. electronics consultant

Of course, there are disadvantages with small firms too. Bright ideas are of little use if the firm has insufficient money to pursue them. One Motorola spin-off, Dickson Electronics, sees its recent takeover by the giant Siemens as being a solution to this particular problem.[67] While it is relatively easy to find many examples of small firms which have been successful in the semiconductor industry, it is much more difficult to trace the fortunes of probably many more which have failed. Between 1954 and 1971, for example, eighty-six firms entered the transistor business, but no less than forty-three are known to have left, the vast majority of these being small companies which never made it.[68] Nevertheless, the impact of some of those companies which did succeed has been considerable. Texas Instruments was a relatively small semiconductor firm when it brought out the silicon transistor, as was Fairchild when it introduced planar. Some small firms which were less than complete successes have also had a powerful influence upon the technology of the industry. General Microelectronics and General Instrument persisted in their attempts to develop MOS against considerable odds.[69] Their progress in this area was the principal factor in the widespread adoption of the technology throughout the industry. By 1968, General Instrument produced over half of all MOS circuits with American Micro-systems (a new firm formed by General Microelectronics men when that company was taken over by Ford[70]) runner up,[71] a situation comparable to that of the late fifties when the two new firms, Texas Instruments and Transitron, dominated the transistor market.

In the fifties, the semiconductor industry was seen as a product of a scientific triumph and vast sums were spent by the larger firms on research laboratories to try to further the progress that had been made. By the late fifties, it had become clear to some that, though the industry was indeed founded upon basic scientific discoveries, its progress was to be determined by the efficiency of its technology. By the early sixties, it had become evident to many more that commercial factors would be the determinants of what technology was useful and what was not. A rift developed between those who sought to advance semiconductor development by pushing forward the frontiers of science and those who sought to achieve the same aim by fully exploiting areas already occupied. New lands discovered by the former group, such as lasers or electroluminescence, are regarded with disdain by those who have spent many difficult years making the semiconductor area economically viable and socially useful.

> There began to be an awareness [c. 1960] that solid state physics had left the laboratory and had begun to be part of the factory. In other words, the engineers and technologists were by and large no longer listening to the basic scientists and really weren't even very much interested in what the basic scientists were doing.
>
> Douglas Warschauer

The early close association between basic science and the commercial semiconductor world seems to have become more distant in the early sixties, when it became apparent to the industry that its future lay in making and selling germanium and silicon semiconductor devices. Many scientists, particularly those working at some distance from industry, were not interested in production processes or marketing and, by the early sixties, often found germanium and silicon overworked. They turned their attention to the more exotic semiconductors, to the III–V and II–VI compounds and to organic materials. Even in the mid fifties, there were about a thousand papers published annually on semiconductors and closely allied subjects[72] and by the early sixties some saw semiconductors as the perfect vehicle for research rather than as materials basic to an important industry.[73] In the development of semiconductors, the transition from the basic science required in the invention of the transistor to the sophisticated technology in the modern electronics industry has reawakened the age-old argument about exactly what it is that science contributes to technology. Not only does the industry feel that its interests are now divorced from those of science, but that it requires a period of consolidation within its existing technology.

Recently I have been shocked to find out that what we were doing to one or two significant figures back in the 1950s is still going on [in universities], but people have three significant figures and are looking at every little bump of the band structure and are just as passionately interested in it.   Douglas Warschauer

In any industrial organisation, the only reason you are doing research and development, quite frankly, is to increase your share of future markets . . . If we are going to work on something to essentially further the state of the art, then we would like to make sure that the output of that is one that would have a significant reward for us in the marketplace.   Mike Callahan

If anything kills the golden goose it's the fact that it's been too successful. There are so many things you can now do with what has been invented that industry and technology can't keep up. They have so much to do to exploit what's been done . . . and there's so much to do in improving this from a technological point of view that there's just no need for spending money on finding new devices.

Ralph Bray

Silicon is it. I see no reason to go in any other direction for a long, long period of time. In fact, I personally killed all the gallium arsenide power diode programmes in Westinghouse. In 1962, we were at a point where ways had to be found to purify the basic materials – too expensive.   William Winter

There hasn't been a whole lot of search for a replacement for silicon for years now . . . I would say rather that there is a wide-eyed recognition in the industry that we haven't begun to apply the technology we have.   Floyd Kvamme

Not surprisingly, the sums spent by the industry on what is loosely described as 'basic' research have declined. As a proportion of sales, research and development expenditure had dropped from 27 per cent in 1958, to perhaps as low as 6 per cent by 1965.[74] In 1972, of about $36 million spent by the semiconductor industry on basic research and development, some $28 million, or about 80 per cent, was spent by Bell Laboratories and IBM, neither of which sells on the open market.[75] In such low esteem is basic scientific research now held in the industry, that many companies, particularly the smaller ones, profess to having little interest in its findings, even when carried out at the expense of others. Relations with researchers in universities are often poor, despite the importance universities are often said to have had in the location of semiconductor companies, and there is a general feeling that the universities now learn more from industry than industry learns from the universities, even

in basic semiconductor research.[76] Not all those working in universities would disagree with this.

> Most of the good basic work in semiconductor device physics is done
> in industry today . . . it's not done in universities.
>
> George Heilmeier

> Some American science professors are opposed to admitting they are
> doing anything useful with practical applications.
>
> Edward Poindexter

Research and development in semiconductors now has to be very
applied if it is to suit the tastes that are now the fashion in the industry. Texas Instruments, for example, no longer uses the term
'basic research' and talks instead of 'total technical effort', a phrase
which embraces a good deal, including even some aspects of
marketing.[77] The industry seems to have found the technology it
wants and that is based firmly on silicon. The feeling is that the
technology will change, but that this change will not arise from basic
research and that nothing is going to displace silicon. This is a situation which some who remember the days of black magic, hit-or-
miss technology of the transistor industry of the fifties find
somewhat disturbing because semiconductors still pose problems
requiring basic scientific solutions which would have been sought 15
years ago. The tendency is now to avoid the problems by finding
alternative ways of doing things or by finding *ad hoc* solutions for
which no explanation is demanded.

> Many people in the semiconductor side of R & D, like myself, are very
> concerned about the areas where we have made conclusions which
> happen to be right conclusions to get some product out the
> door – and maybe it's the best thing since sex, I don't know – but we
> are very concerned because we don't understand what we have done
> to do this . . . We look for 'fixes' to accomplish a solution to a
> problem . . . Most of the good professional people worry about the
> things that they don't understand, and I think that, given the oppor-
> tunity, they will investigate these things.   Kenneth Manchester

> We do not have the ability at the present time to look at the perfor-
> mance of an integrated circuit and say, 'Ah ha! That performance is
> due to the fact that the diffusion was conducted at a degree or so
> higher than it should have been'.   George Heilmeier

> While down at the level of holes and electrons in the active part of the
> semiconductor we have to have a pretty good grasp of what's going
> on . . . it is absolutely commonplace in the structure elsewhere,
> whether in the attachment of the die to the header or in the fabrica-

tion of the package, to have significant technical areas where we don't
really understand what we are doing on any deep theoretical level.
James M. Early

If relations between universities and industry have become less
close, so too have those between the Military and industry. In the fifties and early sixties, the Military was a main market for semiconductor products and often the only one for new and expensive components. New firms often found the military market a particular
boon. Transitron, for example, had little other market for its gold-
bonded diode when it started operations in 1953.[78] That situation
has now changed and the industrial, computer and consumer
markets are either actually or potentially of more importance. One
company with a steady market for 15 million diodes at $3\frac{1}{2}$ cents each
refused to test them to higher specifications for sale to a smaller
market even at $1.50 each.[79] The interests of the Military in high
reliability and performance, whatever the cost, can no longer be
those of the industry if it hopes to compete in these commercial
markets. A good example of this changing situation is plastic encapsulation, for years rejected by the Military, but just recently coming
to be accepted because it is what industry wants to provide for its
other customers.[80] The Military still insists on manual rather than
automatic testing and that none of its components has been
assembled abroad. As other, more valuable, markets are quite
willing to accept these processes, and as the industry is increasingly
unwilling to make special arrangements to satisfy a reduced military
market, it seems likely that the Military too will come to accept these
conditions.[81] If the Military once sponsored devices that eventually
became cheap enough for other markets, those markets are now
buying devices in such huge quantities that their high yields and
consequent high reliability make them of interest to the Military.[82]

There is a division of opinion between those who claim that
military funding of semiconductor development in the fifties was essential to that development and those who see only the market
provided by the Military as having been important. The latter group
suggests that development would have occurred along the same
lines and at the same pace had funds come from other sources.
Whether such copious funds would have been available from other
sources is perhaps less certain, but the argument reflects a feeling
that much military research money was less than fully effective.
Prolonged military support of the anachronistic micromodule

project, unbridled military enthusiasm over molecular electronics, the relative weakness in the commercial market place of some of the larger firms to which the Military gave most support and the outstanding success of such firms as Fairchild, which usually avoided military involvement, are typically used to support this argument.

> With very few exceptions, the major motivation behind technology development cannot come from the Military . . . the major motivation, I feel, is the commercial one . . . I would say that the research that was motivated by getting to a given end result was far more productive than the research that was carried on for the sake of carrying on research. A lot of the Military directly funded research was the latter. I would say that the Military created more motivation for doing good research by creating a market for advanced products . . . The main reason we stayed clear of military involvement was because I thought it was an affront to any research people to say that you are not worth supporting out of real money . . . In a sense, the military funding made whores out of all the research people. You were dealing with a critic of the research you were doing who was not capable of critiquing the work . . . There are very few research directors anywhere in the world who are really adequate to the job . . . and they are not often career officers in the Army.
>
> Robert Noyce, formerly of Fairchild

Despite the success of its product, the American semiconductor industry has faced and continues to face certain serious problems. Many of these are, in fact, closely associated with the success of the product. That success has encouraged both technological improvement, particularly in production processes, and the entry of new firms. Rapidly improving product, improving production techniques and increasing competition have resulted in the price cutting that is the bane of the industry. Companies within the industry have had to learn to live with constant change in order to survive. This is the core problem the industry has to face and its preeminence is the central reason for so very many accusations that, after so many years, the semiconductor industry remains an immature industry.

The rate of change has brought the industry other difficulties. Such a huge industry of thirty years' standing, manufacturing millions of similar or identical components, would normally have long since resorted to automatic methods of production. In the semiconductor industry, this has happened only recently and on a limited scale. Both product and process have changed so rapidly that automatic production lines have generally become obsolete before they have recovered their cost. Probably the best example of this is the automation employed by Philco to make its jet-etched transistor in 1956.[83] By 1963, the Philco technology, once supreme, had become obsolete in the face of the planar technique and Philco, far from having stolen a march on its competitors, departed from the transistor industry.[84] Most companies, especially the larger ones, would dearly love to automate their production lines and many have tried at periods in the past, when they have felt that the technology was becoming stable. Their attempts have nearly always cost them dear and there is, even now, in the very industry which has made possible so much automation in so many other industries, remarkably little automation. Some individual processes, such as testing, are accomplished automatically, but production lines are mechanised rather than automated and the industry remains heavily labour intensive, though the sophistication of the product and productivity have increased enormously.

> Every one of the major manufacturers has been stung at some time or the other with an obsolete technology without exception.
>
> Jerome Kraus
>
> I would guess every semiconductor plant has a warehouse somewhere full of machines that almost worked, but they were a little bit too late.
>
> U.S. semiconductor executive
>
> We don't use many more people to make a microprocessor than we used to use to make a few transistors, so on that basis it's really dramatic how much we have done with so few people.   Gene Strull

Marketing of the industry's products has often been difficult because of the rapid rate of change and severe price competition. It has been well-nigh impossible to make long-term marketing plans without reliable long-term product and price forecasts, and the difficulty has been aggravated by the fact that much of the exploding demand for semiconductors has been not to replace traditional electronics, but in new and often unforeseen applications. Marketing difficulties are probably worsened by the tendency for semiconductor managers to be technical rather than commercial men.[85] The difficulties which arise when the businessman/scientist typical of the semiconductor industry has greater ability in business than in science can be equally severe and in an extreme case result in the successful marketing of a product which cannot be made.

> If the semiconductor managers weren't such idiots, we might not
> have integrated circuits where they are today, because somebody
> would have said, 'Look, I can't manufacture them at the cost I've got
> to sell them'. It was the wrong decision, but it has benefited us all.
>   U.S. semiconductor executive

One consequence of the integrated circuit has been the forging of
new relations between the semiconductor industry and its
customers. This has often taken the form of vertical integration. A
firm that used just to be a systems house assembling components
into a complete piece of equipment has sometimes chosen to make
its own integrated circuits. The computer company, IBM, which
used to buy discrete semiconductors from Texas Instruments, now
manufactures its own integrated circuits and is itself a leading
semiconductor company.[86] ITT in telecommunications, Honeywell
in computers and Cincinnati Milicron, a manufacturer of milling
machines, are examples of firms which had little to do with transistors, entering the integrated circuit business because of the
demands of their own industries.[87] More common has been the
move of semiconductor firms into the equipment business. When so
much assembly had already been achieved in the integrated circuit,
it seemed logical to go one stage further and reap the full profits by
completing the whole assembly. Such upwards vertical integration
has been especially prevalent in the manufacture of calculators and
small computers. Texas Instruments, for example, used to sell integrated circuits to Bowmar to make into calculators, but later com-
peted with Bowmar in selling calculators.[88]

> The integrated circuits made today are probably the most complex
> industrial product made by world industry . . . It is very difficult if not
> impossible for a systems house buying the components to be conver-
> sant with all these things.   Herbert Kleiman

Although some individuals and some companies have made small
and occasionally large fortunes in the American semiconductor industry, their good luck has hardly been typical. Severe competition
generally keeps profit margins slim and as products and processes
have become obsolete, so many of the individuals and companies
which promoted them have fallen by the wayside. There have been
times when the conditions of the early sixties were repeated and the
whole industry has suffered. 1967 was a bad year for semiconductors, but worse was to come.[89] Stock in IBM fell from a 1967–8 high
of 375 to a 1970 low of 218 while Texas Instruments' stock fell 57

per cent over the same period and that of Fairchild no less than 82 per cent.[90] Nor has the first half of the seventies been particularly kind to the American semiconductor industry. 1971 was a poor year, 1972 little better, 1973 saw a shortage of silicon and 1974 has been described as 'just a horrible year'. 1975, it seems, was just as bad.[91] But because so much is now dependent on the semiconductor product and because so much more could be, the industry is optimistic about immediate and long-term prospects. That optimism is founded not only on the industry's confidence in its science or technology or even in the semiconductor product, but also a confidence in the industry's growing understanding of the role of high technology in commerce.

CHAPTER 10

International diffusion of semiconductor technology

In the production of semiconductor components, the rest of the world has always followed where the United States has led. Ever since the invention of the transistor, nearly all major semiconductor innovations have originated in the United States and have been adopted later by other countries. The delay in adoption is known as 'technology lag' and has been traced for eight major semiconductor innovations of the fifties and five of the sixties. The results appear in Table 10.1 and show two features of interest. Firstly, there has

Table 10.1. *Average international technology lag in semiconductors (years)*

	U.S.	Britain	France	Germany	Japan
8 innovations in the fifties	0.1	2.6	3.0	2.4	3.4
5 innovations in the sixties	0.0	1.6	2.6	3.0	1.2
13 major innovations	0.1	2.2	2.8	2.7	2.5

Source: John Tilton, *International Diffusion of Technology,*[2] pp. 25–7.

generally been a delay of at least two years between first successful commercial production in the United States and first production in Europe and Japan. Secondly, though there is still a significant lag, it was, in Britain and Japan especially, considerably shorter in the sixties than it had been in the fifties.

Some technology lag is, of course, inevitable, but it would seem that it is greater between the United States and other countries than between an innovation leader in the United States and the companies in that country which choose to follow its example. Fairchild started commercial production of semiconductor components by the planar process in March 1960, and within twenty months four companies were using the process commercially. It was again Fairchild which introduced the planar diffused integrated circuit in August 1961 and was copied by four other companies within seventeen months.[1] Given the importance of personal contact in the diffusion of semiconductor technology among American companies and the consequent significance of such factors as geographical proximity, it is hardly surprising that a promising new technology is more rapidly taken up by American firms than by foreign companies.

Yet, the penetration by semiconductors of the active component market was as substantial by the mid sixties in Britain, France, Germany and Japan as it was in the United States.[2] By this date at least, the technology lag was not matched by any equivalent lag in demand. Consumption of semiconductors by these four countries, once dwarfed by American consumption, has been growing at a faster rate since the mid fifties than semiconductor consumption in the United States. Table 10.2 shows how great has been the growth in semiconductor usage outside the United States. But the rise in consumption and the apparent closing of the technology gap are not generally evidence of a developing and prospering non-American semiconductor industry. The world semiconductor industry is probably dominated more now by the American industry than it ever has been. For example, U.S. exports of semiconductors accounted for about 11 per cent of consumption in Britain, France, Germany and Japan in 1960. During the late sixties, that figure approached 40 per cent.[3] The last chapter explored some of the characteristics which seem responsible for much of the progress of the American industry; this chapter, while focusing on Britain, seeks to discover why the rest of the world lags so far behind the United States.[4]

The American semiconductor industry exerts its influence on the rest of the world in several ways. Firstly, and most obviously, it exports its semiconductor products. American exports of semiconductors, nominal in the fifties, grew to be worth about $80 million by 1965 and about $470 million by 1972.[5] The huge increase to $848 million in 1973, while it confirms the trend, may not be strictly comparable because of different methods of counting used in that year.[6] No doubt inflation further distorts the figures. There are several reasons which explain the late penetration by the Americans of foreign markets. As Table 10.2 shows, until the mid sixties there was not much of a foreign market – by American standards – to tempt participation. What market there was did not particularly appeal to the Americans. At this period they produced high quality semiconductors firstly for the Military and only secondly for the industrial

Table 10.2. *Consumption of semiconductor devices ($M) (uncorrected values)*

	1956	1960	1965	1970	1972
Britain	2	28	72		210
France	2	27	67	420	114
Germany	3	25	52		218
Japan	5	54	132	420	742
Total	12	134	323	840	1284
United States	80	560	1064	1547	1708
Ratio of foreign to U.S. consumption	15%	24%	30%	54%	75%

Source: William Finan, *The International Transfer of Semiconductor Technology*,[3] p. 94.

and consumer markets. But in most foreign countries, there was no important military market, certainly nothing comparable with the American one, and in Germany and Japan, none at all. Chief demand for semiconductors came from the industrial and consumer sectors and these were very much more concerned with price than was the Military. Those few American semiconductor companies which did export before the early sixties generally did so because European prices had not been as far depressed as those in the United States.[7] For example, a Texas Instruments transistor introduced to both Britain and the United States in early 1961 sold for $14 in the United States and for $28 in Britain.[8] The fierce price-cutting which had been a feature of the American market in the late fifties and early sixties did not occur abroad until the mid sixties. Only when foreign prices began to decline and sales to increase at the same sort of rate as American ones, were American firms interested enough in the European market to be troubled by foreign import quotas and tariff barriers. By this time, of course, not only had foreign consumption reached levels which interested the Americans, but the primary American market had also changed. By the second half of the sixties, the industrial and consumer markets had assumed primary importance in the United States and the semiconductor industry chose to cater for their demands for large quantities and low prices. Identical demands had long been predominant abroad and were rapidly becoming worth filling.[9]

There were various ways, apart from exporting directly, by which American companies could participate in the foreign semiconductor market. The licensing of foreign companies to use American patents was one way and this has been adopted by many American firms. Both Texas Instruments and Fairchild have about half their patent licences issued to foreign firms and 65 of Western Electric's (Bell Laboratories) 179 licensees in 1974 were foreign.[10] American firms earn considerable sums from the resulting royalties. Fairchild's royalty income in 1971 was $9 million and the Japanese paid over $25 million to American semiconductor licensers in 1970.[11] As in the United States, patent licences are often only a formal acknowledgement that one firm has been infringing another's patent rights, and while royalties are no doubt welcome, of much more importance is access to the patent rights and technological capability of the licensee. Occasionally the transfer of technology is aided by agreements by American firms to allow foreign companies to second source their products, and 'know-how' licences as opposed to patent licences. Both involve instruction in process technology rather than simply product design and are, therefore, more likely to stimulate successful innovation.[12] Both, however, are rare. A survey of forty-two American semiconductor firms revealed that only four companies regularly divulged process technology to foreign firms.[13] It appears that licensing policy has generally been defensive or that licences have been used to get extra return on R & D expenditure. They have not been used as a ready means of playing a part in the foreign semiconductor market.

There has been considerable energy shown by American semiconductor companies in setting up their own factories in other countries. In the developed world, these are of two types. A complete semiconductor plant may be built to carry out the whole manufacturing operation and to sell to the local market. Alternatively, the new factory might simply assemble partly completed components, sent from the United States, before selling them locally. Table 10.3 gives an idea of the growing role played by these two sorts of factories in Britain, France, Germany and Japan. The late sixties witnessed an enormous increase in such American operations, which no doubt compensated for the 50 per cent reduction between 1969 and 1972 in the proportion of the semiconductor consumption of these four countries supplied by direct exports from the United States. Between 1968 and 1972, direct exports to these four countries

Table 10.3. *United States participation in foreign semiconductor markets*

	Semiconductor consumption in Britain, France, Germany and Japan (uncorrected values) ($M)	Percentage supplied by direct exports from U.S.	Cumulative number of U.S. factories assembling in these countries	Cumulative number of U.S. factories fabricating in these countries
1960	134	11	5	4
1961	151	15	6	4
1962	174	16	6	4
1963	208	17	7	5
1964	248	16	8	5
1965	323	23	8	6
1966	349	27	10	7
1967	390	32	11	7
1968	490	30	13	8
1969	660	37	24	15
1970	840	30	29	16
1971	875	25	30	18
1972	1284	18	34	18

Source: William Finan, *The International Transfer of Semiconductor Technology,*[3] p. 116.

(which, in 1972, absorbed three-quarters of the non-communist world semiconductor production, excluding the United States) increased 102 per cent while the value of the produce of assembly and fabrication plants rose 358 per cent.[14] The result has been the increasing domination by the Americans of the semiconductor market of these four countries shown in Table 10.4, and obtained not so much by an increase in direct exports, as by the growth of American manufacturing activity in these countries.

An even greater explosion in the manufacturing of semiconductors by American firms has taken place in the developing countries. In 1974, a sample of thirty-two semiconductor firms in the United States, representing 75 per cent of that country's production, controlled thirty-nine assembly and manufacturing plants in developed countries, but sixty-nine assembly plants in underdeveloped countries.[15] Such countries do not participate in the complete manufacture of semiconductor devices, but only in their assembly. The

Table 10.4. *United States share of foreign semiconductor markets (per cent)*

	1968	1972
Britain	53.0	58.6
France	33.0	95.0
Germany	36.0	51.4
Japan	10.0	12.4

Sources: John Tilton, *International Diffusion of Technology,*[2] pp. 115, 144; William Finan, *The International Transfer of Semiconductor Technology,*[3] p. 118.

reasons for this are not hard to find. As we have seen, the semiconductor industry in the United States has always been highly labour intensive with manual and mechanical production being far more typical than automation. Semiconductors being tiny and easily transportable, it made sense to export the half-finished product for assembly in a country where labour was much cheaper than in the United States and then to import the finished product for final testing. In 1970, hourly earnings for semiconductor assembly in the United States were about four times those of Ireland or Jamaica and ten times those current in most of the Far East.[16] Well over 80 per cent of American semiconductor companies have offshore assembly plants, the bulk of these being in Latin America, particularly Mexico and the Caribbean, and in the Far East, especially Korea, Hong Kong, Singapore and Malaysia.[17] The search for cheap labour even led Fairchild to open a plant on an Indian Reservation in New Mexico.[18]

Fairchild was the first American semiconductor manufacturer to set up an offshore assembly plant. That was in 1963. Since then, the industry has invested something like $450 million into establishing such plants. As with the growth in the numbers of fabricating plants in developed countries, the main expansion of assembly factories in underdeveloped countries seems to have occurred in 1969 and 1970. In those two years, no less than twenty-five of the sample of sixty-nine assembly plants were established.[19] By 1972, there were some 89 000 people employed in such offshore assembly plants compared with about 85 000 employed in semiconductor manufacturing in the United States.[20]

While governments of underdeveloped countries are generally keen to encourage a modern new industry by such incentives as tax concessions,[21] none of this international semiconductor activity would be feasible were it not for the existence of Items 806.30 and 807.00 of the Tariff Schedules of the United States. In brief, these allow the importation to the United States of semiconductors for further processing either duty-free or with duty charged only on the value added by foreign labour.[22] Most American semiconductor firms see the availability of cheap trainable labour as the overwhelming advantage of offshore assembly plants and the greatest disadvantages are imagined to be the risks of political unrest or disruption to transport and communication links.[23] It has been imagined that the location of such offshore activities might lead to long-term benefits for underdeveloped countries in that wafer fabrication, joint ventures and full-scale electronic industries may be established.[24] This is a view which is hard to share. It seems more realistic to see the move offshore of the American semiconductor industry as just one further step along a path that has now been followed for some time. Despite the advanced and rapidly changing technology of the industry, success is a product not so much of high quality, but of low price. Offshore industry permits lower costs and, therefore, lower prices. Once the first American company decided to lower costs in this way, its competitors had to follow. Essential as low labour costs are to the existing American semiconductor industry, other factors of the kind responsible for the prominence of Silicon Valley would be entirely lacking for the ambitious electronics firm in, say, Korea. If the sophisticated European semiconductor industry has failed to compete with the Americans, it would be naive to assume that a completely new industry in an underdeveloped country with the single, and possibly temporary, advantage of cheap labour would be any more successful.

Why, then, has the European semiconductor industry, in particular the British semiconductor industry, fared so badly in competition with the Americans? There is, of course, no simple, single answer to such a question, but rather a number of factors which seem to be relevant. Some of the most obvious factors, however, seem to have been of least significance. It can hardly be crucial that the Americans were first in the field as they could hardly have retained their lead for that reason alone, for many of the very firms once holding that lead have since not only lost it, but have dropped out of the business altogether. Nor can it be of overwhelming importance that the market for semiconductors in other countries is so much smaller than that in the United States. Most American firms have been content to cling to only part of the national market for as long as possible and no firm has been able to base its production on the comfortable assurance of a national monopoly. The combined markets in Japan, Germany, France and Britain now rival that in the United States, but, as we have seen, as these markets have grown, so have the various forms of American penetration of them. A reasonable assumption must be that had these countries had larger markets earlier, American domination would simply have been less delayed.

It is, of course, relevant that these other countries have been deprived of the vast American military market. If it is accepted that this market was crucial for the sale of items on the early and very expensive part of the learning curve, then the absence of such a market may go a long way towards explaining poor performance further down the learning curve. Table 10.5 gives some idea of the relative insignificance of the military semiconductor market outside the United States, and it also shows another factor which has been touched on before, that is the declining importance of the military market in the United States itself. While it may be argued that this market was essential in the early days of the semiconductor industry, it is now no longer reasonable to see a large military market as being essential to such an industry. It is also striking that the Japanese, certainly not the least successful competitors in this industry, have absolutely no military market at all.

One of the most striking features of the foreign semiconductor industry is its similarity not to the modern American semiconductor

Table 10.5. *Military and non-military semiconductor markets (percentage)*

		Military markets	Non-military markets
United States	1960	50	50
	1966	30	70
	1972	24	76
Western Europe	1972	14	86
Japan	1972	–	100

Source: William Finan, *The International Transfer of Semiconductor Technology,*[3] p. 93.

industry, but to that which it was imagined would emerge. In the early fifties it was thought that the established valve companies would be first to master semiconductors and would maintain their ascendancy. In the United States, such firms were overtaken by new companies, but that has not been the case overseas. In Britain in 1968, valve companies still held 32 per cent of the semiconductor market, in France 42 per cent, in Germany 66 per cent and in Japan no less than 72 per cent.[25] As in the United States in the fifties, such firms entered the semiconductor industry early, nearly all being in production by 1954. As in the United States in the fifties, such firms are well ahead of both new firms and foreign subsidiaries in the numbers of patents awarded and in their introduction of major innovations, presumably because of the greater amount of money spent on semiconductor R & D. That this situation has continued in Europe and Japan while it has changed utterly in the United States is due to the activities of new firms in the United States and the virtual absence of that activity elsewhere. In Britain, new firms controlled about 16 per cent of the semiconductor market in 1968, in France 10 per cent, in Germany 2 per cent and in Japan about 18 per cent. As a group such new firms are rare in these countries compared with the American experience and it is virtually unknown for them to develop into giants of the sort the United States has seen.[26] As it seems that much of the pace of semiconductor development in the United States can be explained in terms of the decline of the old and the rise of the new, the fact that this has generally not happened abroad may have some bearing on the apparent lack of success of native semiconductor industries. Why is it that foreign firms have behaved so very differently from American companies in the face of the same semiconductor challenge?

> In this company, which is big, the sailing ship effect is very large. The production division will carry on doing what they have done and do it a little better because they have large technological resources and if you come with a new invention they will be against it. These fellows will say, 'Oh well, we can probably do it by tweeking our existing processes'. And they do that until they make a better transistor than you can make with a revolutionary idea, because in the beginning you aren't high up on the learning curve. So you don't win and it takes a long time, until everybody is really screaming, that the new will take over. British Industrial Scientist

It is possible that foreign subsidiaries of American companies act as a barrier to the flow of innovation from the United States to foreign competitors. American companies with valuable new techniques naturally wish to make these available to their own subsidiaries before coming to licensing agreements with foreign competitors; hence the subsidiary benefits from early knowledge and from the various cross-licensing arrangements of the parent firm. When the learning curve is as steep as it is in the semiconductor business, even a small time advantage can be crucial. Yet, this factor ignores two important points. The first of these is that American subsidiaries have been a relatively late phenomenon. The first in Britain – that of Texas Instruments at Bedford[27] – did not open until 1957 and most were very much later. Before this date, licensing had to be with foreign firms and the chief dispenser of licences at this time – Western Electric for Bell Laboratories – was only too keen to spread its knowledge to any company that wanted it. The high attendance of foreign firms at the important Bell symposia of 1952 and 1956 is an indication of how readily accessible the latest semiconductor technology then was. The suggestion is that foreign firms failed to do as much with this technology as their American competitors. This leads on to the second point. Why has so much of the latest semiconductor technology been almost exclusively American?

Part of the answer – and only part – can be supplied at the most simple level. An industry with an early start, catering for the greatest domestic market and supported by the largest military expenditure is likely to be the first to explore promising new technology. Yet, this is hardly sufficient explanation for its introduction of virtually all important innovations in the semiconductor field. The many new companies operating in the American industry have proved outstandingly successful in manufacturing and marketing innovations, even if these were often based, at least in the early days, on research and development work done by the older companies. We have already noted both the massive contribution of new firms in the American industry and their comparative rarity in other countries. This is important, but it is not merely a cause of different conditions, but also a consequence of them.

There are few new semiconductor firms in Europe because few semiconductor experts have seen fit to break away from established companies to start their own or to join new ventures. Consequently, the meagre sources of risk capital in Europe have never become accustomed to the peculiar demands of the semiconductor industry and have been reluctant to supply capital to finance what entrepreneurs there were. Indeed, new semiconductor companies in the United States would be more likely to attract European risk capital than new European companies, because the former have shown that their kind can succeed and the latter have not. In the same way, those who sought to break away from established companies in Europe often found themselves attracted to the United States by superior opportunities, stock options and high salaries. The numbers of Europeans in the present American semiconductor industry attest to the importance of the Brain Drain in the semiconductor field.[28] The owners of Transitron, for example, used to make an annual recruiting pilgrimage to Europe in the late fifties.[29] But it would be a mistake to imagine the whole European semiconductor field to be packed full of frustrated semiconductor experts with nowhere to go, save the United States. There are other reasons for the low mobility of manpower in the European industry. It is not generally the policy of most European companies to recruit from competitors. It is much more common to hire young men straight from university and to promote from within the company. This is the very opposite of the American situation.[30]

> We don't recruit [men from other companies]. We don't look for them. We recruit in general ... young people because if you recruit people who are already advanced then you block the promotion prospects of your own laboratory. It's difficult enough as it is to get on in research so you don't block their progress by recruiting somebody – only in exceptional circumstances ... In the States they are much more callous about this. They will tell a man to go. We have never told a man to leave.   British Industrial Scientist

Two examples might help to illustrate the situation of some of those working in the British semiconductor industry. Ian Gunn was working for Elliott Brothers in the early fifties, but in utter isolation from other solid state physicists. To break this isolation, he asked if he might occasionally visit the Government's Telecommunications Research Establishment at Malvern. This he did and the then Head of the Semiconductor Research Group describes him as a stimulating character. He was awarded a Research Fellowship at TRE from 1953 until 1956, during which time it became clear that he was not only stimulating, but also aggressively ambitious. His interviewer for a job with GEC had been left with the impression that it had been Gunn doing the interviewing.[31] Gunn was eventually lured to North America, first to the University of British Columbia, by a recruiting officer whose own forcefulness can perhaps be gauged by the fact that he burst into the office of the Head of the TRE Semiconductor Research Group, unannounced and without knocking, and demanded to interview the staff. Gunn later moved to IBM and there invented the Gunn Diode, a cheap means of producing microwaves and a major semiconductor innovation.[32]

The second example concerns G. W. A. Dummer, the man who, in Washington in 1952, first proclaimed the idea of an integrated circuit. Dummer was also employed at TRE, in the Component Testing Group where he was the only scientist in a group of engineers engaged in checking that components met military specifications. The work was somewhat routine and the Group, labelled the 'Dumpire', was not held in high esteem. In the early fifties, the Group's interest in transistors for miniature radar sets was crushed because silicon was available only to the Physics Division. Dummer was personally and virtually alone responsible for the world's first integrated circuit model, displayed at TRE in 1957. Yet, Dummer encountered only delay and apathy when he sought money from the Government or Industry to develop integrated circuits. In 1957, only Texas Instruments, who eventually patented Kilby's integrated circuit in 1959, were enthusiastic about Dummer's model. The work of Dummer's Group received no real backing until the early sixties, by which time Dummer himself had decided to retire to write books on concepts and techniques.[33] The total resources available in Britain or elsewhere in Europe are so much smaller than in the U.S.; consequently a smaller number of projects receives backing. The problem of picking winners is a notoriously difficult one.

> As individuals, European scientists and engineers usually tend to stay abreast very well of the changes in technology both here and in Europe.   Mike Callahan

> I think that the reason why some of these things don't work in Europe is because they never have the resources to produce this wastefulness which makes our system go.   John Hallowes

> In Britain, you can't afford to make a mistake. The only way you are going to progress is if you can afford to make mistakes. If you have one of ten winners, you are ahead in the game. The British can't afford to do that: the Americans can.   Rudolph Verderber

In the United States, those engaged in industry are held in as much esteem as those working in universities or government research laboratories. In Europe this is definitely not the case.[34] There the worker in industry – whatever his rank – is often seen as a much more lowly creature than the scientist working for the government or the scholar in a university. This does not, of course, mean that European industry is entirely composed of third class men, but it does mean that self-improvement, if it cannot be arranged within the confines of the company, is seen in terms of a move to government service or to university rather than to another company. But flow of experts in even this direction is the exception rather than the rule. Various cultural and institutional factors conspire to prevent mobility. A man who has changed his job too frequently or who does not have the recommendation of his last employer may well find that prospective employers regard him with suspicion. Job security is often highly valued, perhaps more than the job itself and such factors as pension plans may make transfer from one job or sector to another financially hazardous.[35] Consequently, there is little mobility among European semiconductor experts and, therefore, few new firms. Because there are few new firms, mobility is not encouraged.

Government policy hardly seems to have helped to remedy this situation. Because semiconductors are both an important modern industry and one upon which many other activities, including defence, are dependent, European governments have shown some interest in trying to improve the lot of their native semiconductor industries. In general terms, they have attempted to do this by supplying R & D funds and by rationalising the industry. Yet, while the American government has chosen to fund the bulk of semiconductor R & D through contracts with industry, the French and British governments have chosen to spend more money in government research laboratories and on university work than in industry.[36] Considering the difficulties the Americans perceive in the transfer of innovation from government laboratories or universities to industry, even when there is some mobility of staff among them, this seems an inefficient policy for countries where there is much less mobility. Nor are the sums involved so trivial as to make their expenditure unimportant. In 1968, the combined total government expenditure of France and Britain on semiconductor research and development probably rivalled that of the American government.[37]

> I would generally say that if you are worried about commercial exploitation of new scientific knowledge, you want to get it as close to the end use as possible, that is go ahead and subsidise company research.   John Tilton

The rationalisation policy of European governments is also directly contrary to the American experience. Both the French and the British governments have seen fit to encourage the amalgamation of existing companies to form giant national ventures, such as GEC-English Electric and Sescosem, with which to answer the American challenge. Government contracts have generally gone to such firms. In contrast, the American government, though supporting the largest firms in the early days of semiconductors, has since been keen to disperse its patronage and indeed, by means of anti-trust legislation, to challenge the growing supremacy of any one firm. While it would be innocent to assume that the American way was the only or even the best way to make semiconductors or that an imitation of it would naturally be the best means of competing with the Americans, countries such as Britain or France do not seem to have found a good alternative.[38]

> [The American semiconductor industry] has been more or less an industry left to its own devices . . . We have not tried to spoon-feed it as you have in the U.K. or the Japanese organise the hell out of it.
>   U.S. expert on semiconductor commerce

> The real failure of the British and the Europeans has been in the technology, not in the science . . . Japan is an enigma to me. It seems to violate all the rules and succeed remarkably well.   John Tilton

> The English took a scientist's route and the Americans took a technologist's route. They spent all this time – and so did the Dutch – worrying about these III–V compounds like gallium arsenide. The U.K. has probably developed more original technology than any other country, besides the U.S., but it has very little to show for it . . . I think for the U.K. industry to become a viable competitor of the U.S. would be very difficult because it's so late in the game and we are so big. Japan, with its more organised, practical approach, conceivably could.   U.S. expert on semiconductor commerce

The situation in Japan is totally different and the chief reason for this appears to be the comprehensive and successful control the government has exercised over the industry. Until 1974, subsidiaries of foreign companies were nearly always forbidden, nor were foreign companies generally permitted to acquire controlling interests in Japanese firms. The government has also exerted considerable influence on the semiconductor industry by controlling licensing arrangements made between Japanese and foreign firms, by attempting to spread the findings of R & D throughout the industry, and by attempting to achieve a balance between the benefits of industry consolidation and those of competition. The result has been the spectacular growth of the Japanese semiconductor industry – faster than those of Europe or even the United States. Restrictions on imports and the absence of foreign subsidiaries have meant that the local market remained the domain of native manufacturers. Yet the local market has never included the military market, which is often seen as having been crucial in American semiconductor development and the paucity of which is generally imagined to have been instrumental in the poor performance of the European semiconductor industry. Instead, the Japanese have always concentrated on the consumer and computer sectors of the market – IBM is one of the few foreign-owned firms in the country – and especially on the export of these products. In the late fifties, for example, two-thirds of Japanese transistor production went into radios and by 1968 90 per cent of these radios were being exported.[39]

In Japan, unlike either Europe or the United States, the established valve firms dominate the market, but not to the exclusion of several new firms. It seems that the older firms in Japan are quite as quick to adopt new technologies as the newer companies.[40] Certainly they have access to technological developments through the same channels, by government dissemination of the results of Japanese R & D or, and this is by far the main source, by foreign licensing, not just of individual firms, but often of the whole industry. Thanks to a liberal licensing policy of the American industry, Japanese firms had rapid access to the latest technology. Their own R & D efforts are very much geared to production improvements and there is little interest in anything resembling basic research. A great deal of Japanese semiconductor production is exported, much already incorporated into electronic equipment and much to the United States. Although Japan may continue to make production improvements, it is rapidly losing the advantage of cheap labour which originally made its exporting attractive. American firms using offshore assembly can generally avail themselves of much cheaper labour. The Japanese industry has by means of established valve firms, extensive government interference, with little R & D, virtually no personnel mobility and with no military market, achieved a measure of success that has been denied to the European semiconductor industry.[41]

There are certainly some observers who view the decision of the Japanese Government in 1974 to allow American-owned companies to operate in Japan as an important watershed. American firms, particularly Texas Instruments, had been showing a growing reluctance to continue a one-way flow of information.[42] Now that American semiconductor companies do operate in Japan, there seems little to prevent them creating the sort of situation that has arisen in Europe, where the latest and most valuable technology is confined to American subsidiaries and is made available to native competitors only when it has outlived much of its profitability. If this does happen, it means that the Japanese industry will face much the same problems as the European one.

It is hardly profitable to adopt the point of view that if only the semiconductor industry in Europe had been more like that in the United States over the last twenty or thirty years, its present problems would not have arisen. Certainly, one cannot help feeling that if more new firms had been active in the field, if staff were more mobile in the industry, or if the integrated circuit had been patented in Britain, things would have been different, but these rather hypothetical conclusions do nothing to improve the prevailing situation.

One option now open to non-American industry seems to be specialisation within the semiconductor field. If it is impractical to compete with the Americans across the whole semiconductor spectrum, competition within a chosen area might be a viable alternative. The choice of a rewarding and expanding area would naturally be difficult and crucial, but the advantages to be gained from being able to devote a large proportion of R & D and production activity to this sector might be considerable. The superiority of the British in microwave development and of the Dutch in the development of television tubes suggests what can be done in the face of strong American and other competition. The argument that such a policy would surrender the remaining market to the Americans is no longer very impressive for the Americans already hold most of this market, especially that for the latest and most profitable devices. Though their possession of so much of this market may be galling for native semiconductor producers, it does mean that the many other industries which use semiconductors have rapid access to the newest products.[43]

A second course concentrates on the advantage the Americans are seen to have in their large domestic and military market for semiconductors. A potential single market of equal or even greater size is seen to lie in the European Economic Community. Such a unit, uncluttered by national considerations, could offer the sort of resources and market in which the American semiconductor industry developed at such pace.[44] The EEC could presumably provide these conditions, but they do not exist at the moment and it is a matter of conjecture how soon, if ever, they will exist. Chief among the obstacles to the success of such a policy, of course, would be the fact that though the American industry certainly developed under these conditions, it did not do so in an environment of competition with a superior foreign producer, as European companies would be forced to do.

Perhaps the boldest suggestion for the salvation of the European industry is one that concentrates on some of the less tangible factors which have been mentioned as having been crucial to the development of the American semiconductor industry. If, it is argued, the present situation of extensive domination of the international semiconductor industry by the Americans is seen as undesirable by foreign nations, and if it is seen as impractical, impossible or pointless to re-organise foreign industry in imitation of American industry, then an alternative may be more convenient and practical. If so much of the American success may be attributed to new firms, clustering of companies, staff mobility, good communications and the American market, then it could be possible to take advantage of these factors without radically altering the domestic industry. It is argued that if a foreign company or even a foreign country were to supply the considerable risk capital for its own semiconductor firm in such a location as Silicon Valley, it would then be in the same position as any other commercial participant in the American industry. It is imagined that such a firm would rapidly discover what ingredients contributed most to success in the industry, that it would act as a training ground for non-American personnel and that it would be able to funnel information about processes and techniques to the parent firm or home country with the minimum delay and the maximum effectiveness.[45] The suggestion seems an interesting one, not least for the obvious importance it puts on personal factors, and the idea no doubt at least partly explains the recent activities in the United States of such firms as Plessey, Siemens, Philips and ICL.

The Americans seem to have a healthy respect for European semiconductor technology and there are some who see a real threat of it catching up with their own.[46] Where they see least danger is in the use Europeans make of their technology. They seem unanimous in identifying the main problem as being the commercial exploitation of that technology. Most explanations for this centre on the very many differences between the way the American semiconductor industry is organised and the way things are done abroad, but again, the performance and the incentives of individuals are often highlighted. Americans often find it strange that so many in the British semiconductor industry, for example, seem to value job security and a pension above all else. Why, they wonder, are semiconductor salesmen in Europe not the experts in their field that their American counterparts are?[47] Why do people in the British industry not work harder? Most telling of all was the comment of the President of one leading American semiconductor firm. Why, he mused, did he, with all his many contacts in the prosperous American semiconductor industry, know only three semiconductor managers with chauffeur-driven limousines when, with very much poorer knowledge of the small and ailing British industry, he knew at least a dozen British managers who were regularly chauffeur-driven?[48]

It would be a mistake to imagine that the shortcomings of the British semiconductor industry can be entirely, or even mainly, explained by those factors which are most readily quantified. Certainly such factors as the market size and type, the paucity of new firms and the level and direction of R & D support are important, but they by no means explain the whole situation. That must be seen in terms of factors which appear less satisfactory because they are impossible to quantify. Such factors are important not only in their own right, but also in the influence they exert to create conditions which are hostile to the development of a healthy semiconductor industry. It has been suggested, for example, that both the internal organisation and the management of British semiconductor companies is often weak.[49] When communication even within a company is poor, it is hardly to be expected that communication with outside organisations will be any better. British companies have failed to sense the vital importance of the rate of change in both technology and market in the semiconductor business[50] and they have generally been unwilling to take calculated risks.[51] This has also happened in some companies in the United States, but generally with the consequence that these companies have been replaced by others more responsive to – even responsible for – the rate of change. That has only happened in Britain in the sense that American subsidiaries have captured more and more of the market. The result is that Britain and most other countries are on the horns of a dilemma. So much of modern social and economic activity is now dependent on semiconductors that any move to restrict the American supply would cripple the development of a vast range of these other activities. Yet, the more these activities flourish, the more the American semiconductor industry prospers and, with its present organisation, the more the European semiconductor industry will, in comparison, languish.

> I frankly don't know what the answer is. I am rather discouraged about how we in the U.K. can get into the market . . . There are certain quarters of the integrated circuit market that we are not doing too badly in, but these are very very small segments of it and we arrived at them largely by accident.   U.K. research manager

CHAPTER 11

Reflections on an electronic age

All generalisations are dangerous, but without them the intellect is starved. The risk, such as it is, is well worth taking. In trying to obtain a general view of the series of interlinked innovations we have described, a pattern seems to emerge which may have some general validity. In the early stages of the innovation, a 'market-pull' situation seemed to exist, in which a steady market for any practical device could be anticipated. Wireless telegraphy could clearly use detectors and, perceived a little later, amplifiers of electrical signals. Later still, when the valve was fully established, a clear market existed for a device which would avoid the drawbacks of the valve without introducing substantial disadvantages of its own.

But even in describing the 'market-pull' for semiconductor detectors and amplifiers, we can see that needs cannot be perceived in complete isolation from invention. Needs and wants cannot be articulated with any precision unless some means of fulfilling them have been demonstrated as feasible by science or technology. Flights of fancy and age-old dreams give some, albeit few, genuine market clues. Before the invention of the aeroplane, a desire for travel through the air had been clearly and repeatedly articulated. In the same way, a desire for free or very cheap mechanical motion, some form of perpetual motion machine, is old, venerable and unfulfilled. Communication over long distances, preservation of perishable foods and perhaps a very few other general desires would complete a very short list for articulated but vague technological desires. To obtain a very, very much longer list of market desires and to obtain anything reasonably clearly defined, desires must be coupled to possibilities. Possibilities mean inventions and then, of course, the border between 'market-pull' and 'invention-push' becomes blurred. The two concepts may be sufficiently separable to serve a useful intellectual purpose, but a really clear distinction is rare.

If the proverbial savage, the person without experience of technology or its products, were taken to a large but empty shop, there is no way in which he could conjure up in his mind all the thousands of different items that normally fill the shelves. If he were asked to imagine all the goods he would wish to have, he would fill no more than a tiny corner of the imaginary store. Even if we were to replace the savage by a person of technological experience and were to ask him to furnish the empty store with goods of his fancy, the result would inevitably closely resemble a store of his experience. Desire is stimulated by the display of wares; unstimulated desire is modest and its inventiveness cannot possibly match the cumulative inventions of thousands of inventors over many generations of civilisation. Even the most sophisticated shopper can do little more than exercise choice between goods. Most of the time he can choose only between wares offered; very rarely can he articulate a truly new requirement. 'We give them what they want' is the emptiest of phrases in a marketing world so rich in empty phrases.

There is one group of wants that can be expressed more accurately, although not sufficiently accurately to produce a shopping list. We may give this group the generic title 'protection from natural afflictions'. This clearly covers protection from, and cure of disease, and protection from hunger, cold, inclement weather, predators and other enemies. Not a long list, and not one which defines clear market desires, but a list of basic human needs whose fulfilment is the most essential task of technology and its first priority in an idealised state of affairs. A fundamental difficulty of articulation of wants arises in this category as much as in any other: the possibility that a technology developed for one purpose will serve quite another. Diagnostic and therapeutic X-rays were a by-product of investigations in pure physics; much modern medical technology requires solid state electronics. The paths of technology are inscrutable and certainly rather convoluted.

After this lengthy digression, we must return to semiconductors. Their history certainly starts, despite all the reservations we have discussed, with a healthy amount of 'market-pull'. But the market would have pulled in vain, had it not been for a sound scientific base to the invention. More than most innovations, the transistor was born out of scientific discovery. No doubt the science was aided by a whole gamut of techniques and instruments, but these served as the tools of science. Many innovations are based on technology, often aided by science at many stages. The transistor is one of the supreme examples of an invention truly based on science.

Even at the scientific stage entrepreneurs had a role to play, but at this stage they were scientific and not commercial entrepreneurs. They were people with vision who did the inventing, people with vision who encouraged and supported the inventors, leaders who kept the scientific teams going and pushed steadily towards the goal. Competition too had a role to play at this stage. One of the main stakes in the scientific game is prestige; the prestige that comes from being the first to articulate a new idea, find a new formula or get a new experiment working. And what better experiment than one that produces a highly desirable technological possibility! The will to succeed, the desire for prestige, the urge to beat the rival are all powerful incentives for scientific effort and certainly played a part in producing the transistor. But scientific management also played a crucial role. The research directors of Bell Laboratories encouraged the search, allowed it to range widely and gave it adequate financial backing. They recruited the best available brains and gave them adequate technical support, and they created an atmosphere of scientific interchange and stimulation. The role of 'product champion' was thus divided between people like Shockley and people like Kelly; and it is an important advantage for any innovation to have champions in powerful positions.

As soon as the transistor was announced, technology took over. Once the invention was made, what mattered most was the ability to make the devices reliably and at reasonable cost. The race was on for manufacturing technology, and a highly hazardous game it proved to be. Invest too little or in the wrong technique and you were out of the game. Invest too much, and you risked hanging on to a process well into its obsolescence which could also see you off. To win in this game required the right manufacturing technique and a good eye for the market. In fact 'market-coupling' was crucial at this stage. The devices were not strong enough to create markets for themselves easily; they had to seek out corners in which they could sell, despite their weaknesses. It was not enough to be able to make devices; they had to be of a kind that somebody could be persuaded to buy.

By rapid exchange of information and feverish activity by technological and commercial entrepreneurs in a generally booming economic climate, the technology of transistor manufacture developed rapidly. There was still close co-operation with scientists, but they were no longer in the driving seat; that was firmly occupied by technologists and businessmen. As the technology developed, the product became stronger and became able not just to seek small corners of the electronic market, but to create markets for itself. This phase became particularly prominent with the advent of the planar process and culminated with the integrated circuit.

From the early 'market-pull' and scientific invention, the scene had shifted to one strongly dominated by technology and the ability of the product to create its own markets. The innovation had entered a phase of 'technology-push'. So strong is this trend that not only can integrated circuits create markets for themselves, they can also alter technologies in related fields. The devices now made are largely dictated by the ease of manufacture, and ease of manufacture brings down their price. Because active semiconductor devices are incredibly cheap in terms of cost per active circuit component, it has become worthwhile to adopt circuits which use more active components and fewer passive ones. Thus we obtain a shift from analogue to digital devices and even telephone transmission is moving towards the digital form. The manufacturing technology of semiconductor devices is beginning to dominate electronic circuitry and electronic machinery. This is not, perhaps, a case of complete technological imperative, but certainly shows some aspects of this phenomenon. Undoubtedly, technological developments in one sphere have influenced technology in numerous other spheres and have created marketable products previously unheard of and undreamt of.

It would be impossible to review all the products that now incorporate solid state electronics and a few examples will have to suffice. We have chosen to take a particularly close look at calculators because they are representative of several important facets of recent semiconductor development. They are an important part of the industry's venture into a relatively new market area, the consumer market. They are also a fine example of the forces encouraging vertical integration in the industry and of the sort of problems which may arise when those responsible for the technology find they are dealing directly with a new and unfamiliar market. The calculator is also an obvious example of the sheer power of semiconductor technology and of how pervasive this power can be. As people once used to look upon the transistor as a radio, they now look upon the

semiconductor chip as a calculator.[1] Moreover, the calculator illustrates the crucial importance of declining price in the semiconductor industry. Lastly, the calculator provides a useful indication of
the relationship between the American semiconductor industry and
foreign industry, particularly the semiconductor industry of Japan.

> I think the electronic calculator is a beautiful example of the per-
> vasiveness of semiconductor electronics and the things that a vivid
> imagination, aggressive and thoughtful semiconductor management
> and entrepreneurialism in the highest sense in a company like Texas
> Instruments can produce.   Harold Levine, Texas Instruments

The calculator is a computer. The most basic and the cheapest
calculators do little more than add, subtract, multiply and divide,
but there is a hierarchy above the basic model capable of more
elaborate functions. Somewhat more expensive is the scientific
calculator and more expensive still the desk top calculator, often
with its own print-out. At the top of the hierarchy, and almost in-
distinguishable from what is normally regarded as a computer, is
the programmable calculator. Prices range from about $10 for the
cheapest hand-held, four-function calculators to several thousand
dollars for the best programmable machines.[2] Clearly, the top end
of the scale merges into computers proper and it is, therefore, the
bottom end that is most distinctive. Here the calculator has not
merely replaced the slide rule and the adding machine, but has
provided a capability that is utterly beyond the capacity of either.

Mechanical calculating machines were available in the thirties but
at about $1200 – the price of a couple of family cars. By the fifties,
electro-mechanical machines were being made for perhaps half that
sum – the price of quarter of a car. The first electronic calculators,
containing discrete semiconductor components wired to printed circuit boards, were produced in 1963 by a British firm, the Bell Punch
Company. The machine was made under licence in America and in
Japan, where the advantage of cheaper Japanese labour for the hundreds of connections required led to a Japanese domination in the
manufacture of calculators throughout the sixties.[3] Integrated circuitry was, of course, the perfect technology for the calculator, and
MOS – slower but more compact and cheaper than bipolar integration – the most appropriate of the integrated circuit
technologies. By the second half of the sixties, calculators using
MOS integrated circuits were available. Both Texas Instruments and

Mostek had announced a two or three chip MOS calculator by 1967, but these aroused little interest.[4] Because calculators still needed several chips and required significant assembly, the advantage still lay with somewhat cheaper Japanese labour. But the calculator in the sixties was never the mass-produced article that it became in the early seventies as the Japanese increased their imports of American semiconductor chips and their exports of finished calculators.[5] American production of MOS chips for calculators was still so low that by 1971, when American companies started assembling calculators, the Japanese industry found itself starved of its raw material.

The first American company to make calculators was a firm called Universal Data Machines, operating from a warehouse in Chicago. The company bought chips from Texas Instruments and, using cheap immigrant labour from Vietnam and South America, assembled five or six thousand calculators a week for sale through a local department store. Probably the second company to enter what was to become a particularly vicious race was the Canadian firm, Commodore, newly moved from Toronto to Silicon Valley. Commodore also used a Texas Instruments MOS chip, but adopted a technology developed by a component supplier, Bowmar, for making a particularly compact calculator. Bowmar had chosen not to make calculators itself and had found no interest in its technology among the established manufacturers of electro-mechanical calculators. Although the first mass-produced calculators dropped rapidly in price from about $100 in 1971 to $40 or $50 the following year, the profits of these small entrepreneurial companies remained high. A calculator selling for as much as $50 in 1972 had probably been made for between $10 and $15. Consequently, a situation arose whereby component manufacturers with a sophisticated technology were, for comparatively little profit, supplying the capacity for others with little technology at all to make vast profit with minimum effort.[6] The situation had to change as it became staringly obvious where the profits in the exploding new market lay. By 1972 Bowmar was struggling to get back into the business it had earlier farmed out and was joined by other semiconductor manufacturers, including Texas Instruments.[7] The calculator provides perhaps the best example of rapid vertical integration in the semiconductor industry, but if small firms had not demonstrated the viability of the new product, it is doubtful whether such integration would ever have taken place or,

indeed, whether the calculator would ever have gained the acceptance it has. In the late sixties and even the very early seventies, few
in the semiconductor industry anticipated the sort of future that lay
ahead of the calculator.[8]

> I think Texas Instruments and all the other big companies would not
> have gone into calculators had they not seen the success of the little
> guys.   Suhael Ahmed

> I don't think that ten years ago anyone could really have said that the
> hand-held calculator market would get to the proportion it is now
> ... The consumer products market is the great future of the semicon-
> ductor because it has been the most resistant and is the last of the
> great markets to fall.   Jerome Kraus

Some of the largest semiconductor manufacturers simply could
not believe that these small assembly firms were justified in their
orders of massive numbers of components and chose to make their
own, substantially lower, estimates of what the demands of these
firms would be. As it happened, this was sometimes a fortunate decision, not because there was reduced demand for the calculator, but
because many small firms met with cash-flow difficulties and were
unable to pay for components.[9] With the rapid growth of vertical integration, some companies not making their own chips found
themselves in short supply. Bowmar even tried to sue Texas
Instruments when its supply of calculator chips was cut off. The
Japanese fared particularly badly, for not only did they depend on
MOS chip exports from the United States, but they also lacked the
necessary technology for producing the light emitting diodes used in
the calculator display.[10] Yet, by 1975, both components were readily
available to anyone who wanted them and, in true semiconductor
fashion, their price had dropped drastically. Thus, although there
was still little labour required in calculator assembly, the labour cost
had increased significantly as a proportion of component cost to
once again give the Japanese an advantage. The latest development
promises to turn the tables yet again. General Instrument has
recently produced a hybrid module containing MOS chip, display
and the driver which activates the display. The module is sold to
anyone making calculators and requires only a battery and a case,
both of which can be connected automatically. Once again, it seems
the Japanese will lose their advantage.[11]

The rapidly declining price of calculators has been one of their

most striking features and there is apparently still room for further
price reduction.[12] More advanced calculators are much more expensive, but their price has also been declining. Table 11.1 gives an idea

Table 11.1. *Average U.S. hand-held calculator prices ($)*

	Consumer	Scientific	Business
1974	26.25	80.60	174.20
1975	21.00	59.40	56.40
1976	17.50	47.40	44.40
1977	14.00	38.40	35.40
1978	11.20	32.40	29.40
decline			
1974–78	57.3%	59.8%	83.1%

Source: Coleman & Co. estimates, 1975.

of relative average prices of hand-held calculators in the consumer,
scientific and business markets and suggests that the greatest price
reductions will occur in the more elaborate calculators. One scientific calculator brought out in 1975 at $225 had dropped to $150
within two months and to $137.50 during the following one and a
half months.[13] The largest price reductions in the consumer models
have already taken place, much to the discomfiture of some companies in the business. Bowmar, among many others, has recently
elected to leave the business altogether and the main interest in the
industry is participation in the more sophisticated calculator market
while there are still profits to be made there.[14] Table 11.2 shows the
estimated size of the world calculator market for the mid seventies
and is quite clear about where growth is expected.

The sort of price decline that has been and still is evident in the
calculator business, dramatic as it is, should hardly have taken
anyone, especially those in the industry, by surprise. After all, it is
exactly the sort of thing that has been happening with semiconductor components for the last two decades. Once the calculator had a
market of the same magnitude as that enjoyed by most semiconductor components, it was inevitable that exactly the same thing should
happen. Consequently, the basic, four-function, hand-held
calculator has dropped in price from about $100 to about $14
within five years. One semiconductor manager remembered being
very proud of a $200 calculator brought back from Japan in about

Table 11.2. *Estimate of world calculator production and value*

	Consumer		Scientific		Business	
	Units (M)	Value ($M)	Units (M)	Value ($M)	Units (M)	Value ($M)
1974	30.2	820	1.9	168	2.0	255
1975	39.9	881	3.8	248	4.8	418
1976	51.1	922	6.1	317	7.5	562
1977	59.3	859	8.9	375	11.1	713
1978	64.7	758	12.4	440	15.0	866

Source: Coleman & Co. estimates, 1975.

1970, but the pride wore thin as the price dropped to about $15 in 1975 for a calculator providing the same functions.[15] That price decline must now start to level off simply because the cost of electronics is no longer a major part of total costs. This is not yet the case with more sophisticated calculators, but it may well become so. There is also some difficulty matching advanced electronics technology and its low price with other essential but relatively unsophisticated and expensive technology. Plastic cases, for example, provide barely adequate housing and their press buttons are the part of the calculator most likely to fail. Table 11.3 compares the

Table 11.3. *Manufacturing cost of consumer calculator and of scientific calculator ($)*

Consumer calculator		Scientific calculator	
MOS integrated circuit	1.50	MOS integrated circuit	8.50
6 digit LED display	1.50	12 digit LED display	3.20
Display driver	0.50	Display drivers	1.20
Disposable battery	0.25	Rechargeable battery and charger	5.70
Instruction book, warranty card, sleeve, box, etc.	0.16	Printed circuit board and miscellaneous parts	0.50
Keyboard and case	1.20	Keyboard and case	2.90
Labour	0.25	Labour	2.00
Total manufacturing cost	5.36	Total manufacturing cost	24.00

Sources: Nicholas Valéry, 'The electronic slide rule comes of age',[2] p. 511; Coleman & Co. estimates, 1975.

manufacturing cost of a very basic, hand-held calculator with that of a scientific calculator.

The combination of low price and mathematical capability of the calculator has led to its acceptance on a scale undreamt of by even the most optimistic in the industry. By 1975, there were about 35 million calculators of various sorts in use in the United States alone, two-thirds of these having been bought during 1974 and a third received as presents.[16] In Britain, it was estimated that 4½ million would be sold in 1976.[17] Some of the problems this calculator fever has produced are fairly obvious. For instance, the high technology of the semiconductor industry now has to be geared to cater for the annual Christmas spending spree. Some suggest there are more insidious problems. For example, it has been claimed that children, armed with calculators from infancy, are unlikely ever to achieve any real understanding of what it is they are doing with numbers.[18] That may be, but what is more certain is that millions of people who are not scientists or businessmen, that is those who might reasonably be expected to juggle with figures as a major part of their daily toil, now have their own calculators. Most of these people would, presumably, never have dreamt of buying a slide rule or an adding machine, and yet they perceive the need for a calculator. It would be fascinating to know how frequently these are used, for what purposes and how much of the total capability of the calculator is being utilised. How many housewives actually need to know the square root of a number? But then the technology is ridiculously cheap. For a fraction of the cost of one week's housekeeping, one can have permanent access to any number's square root. Perhaps there is a point below which it is enough that technology is cheap and unimportant whether it is particularly useful. Calculators have become a fashion – so much so that manufacturers expect some of their custom to come from impulse buying. The semiconductor industry has obviously come a long way since the days when the Military was its main customer. The technology has advanced and in doing so, it has become more and more pervasive. The industry has reorganised itself to cope with this new sort of mass demand and now looks upon the consumer market as being the most important of all.[19] The calculator has led the way into this promised land, but it remains to be seen what the industry will do there and exactly what the inhabitants will want to have done.

People didn't know they needed hand-held calculators until they were available. Floyd Kvamme

Up until the present time . . . most of the contact that the consumer has had with the semiconductor manufacturers has been very indirect, but I think that is going to change. Mike Callahan

I find that the growth of the calculator and the watch industry is opening up a completely new awareness of the public. The public is now electronics-minded. They are looking for the next gadget to fiddle with. Suhael Ahmed

Tell somebody to divide 12 by 12 and they pull a calculator out.. Suhael Ahmed

An even more recent product of the semiconductor industry than the calculator is the electronic watch. Like the calculator, its manufacture was a gamble in that no one could be certain whether the product would be welcomed. Unlike the calculator, and perhaps as a result of the lesson it taught, the fully electronic watch was first produced and marketed by the large semiconductor manufacturers. Also unlike the calculator, the electronic watch is faced with established competition from the traditional watch industry.

In 1973, there were about a quarter of a million electronic watches produced, virtually all in the United States, and selling for an average price of about $250. By 1975, production had reached 3 million at an average price of about $150. By 1980, it is imagined that world production of electronic watches will have reached 50 million, at an average price of perhaps $40 and that electronic watches will then have captured over 16 per cent of the world watch market and over 40 per cent of the American watch market.[20]

There are, in fact, three main types of electronic watch. All work by using electricity from a tiny battery to make a quartz crystal vibrate. The vibrations produce electrical pulses and these regular pulses are then counted by an integrated circuit which is used to activate the digital time display. The LED (light emitting diode) display is identical to that used in the calculator and, to save power, requires the pressing of a button to tell the time. The LCD (liquid crystal display) uses reflected light to pick out the numbers and often carries its own light source to allow the watch to be used at night. The quartz analogue watch is quite conventional in appearance, using a small motor, controlled by the integrated circuit, to drive round the hands. As yet, it is far from certain which of these three types, or any newer types, will prevail, though with the LED watch priced as low as $35 in 1975, compared with a minimum price of $60 for the quartz analogue and $95 for the LCD, this is perhaps the likely favourite.[21] It seems almost inevitable that the price of electronic watches will follow the established pattern of semiconductor prices in general and those of calculator prices in particular.[22] If this does happen, the most cut-throat price competition will be in the cheaper watches, particularly those with LEDs, and it may be that some manufacturers will endeavour to stay clear of this market by making more expensive watches which do not require the pushing of buttons.

All electronic watches are impressively accurate to within a few seconds in a month. Presumably there is a demand for such accuracy, though it is hard to see what advantage it can be to most people. Electronic watches also require batteries and these must be replaced periodically, though it is likely that this extra expense will be adequately compensated by the less frequent cleaning necessary. It will be particularly interesting to see what relationship is established between the new electronic watch industry and the conventional watch manufacturers, especially the Swiss. Some imagine that the Swiss will be forced to lean more towards the jewellery market, that most cheaper watches will soon be electronic and that the electronic watch will soon dominate even the quality watch market.[23] That is unlikely to happen until electronic watches overcome their display problems and have demonstrated that they are likely to last. Yet, electronic reliability is likely to be a feature of mass production and rapidly declining learning curves. In semiconductor electronics, the other feature of these is, of course, low price. Consequently, there may be a situation in which the cheapest electronic watches are also the best electronic watches.[24] That state of affairs would not only undermine the conventional watch industry, but, if extended to a wide range of other products, would have an interesting impact on prevailing ideas about cheap mass-produced goods and on the general concept of value for money.

> The watch that will be the most reliable will be the cheapest because it will be the one that's produced in largest quantity and we will have most of the bugs out of it and will have more field experience with it to see what can go wrong, and we will have corrected it.
>
> Robert Noyce

> If one has to fight an old-established market without having something that's completely new and serves a new function, then you at least economically have a problem . . . No one seems to have really sat down to think what other types of functions could these watches serve or could a personal electronic gadget serve other than as a watch.
>
> William Corak

Telecommunications is one field in which there is immense scope for semiconductor electronics to affect popular lifestyle. We have already seen that radio, in its use of cat's whisker crystals, was one of the earliest applications of semiconductor devices. This exploitation has continued, though now it takes such forms as communications satellites and tiny television cameras. The telephone, although forming only a fraction of the total telecommunications system, is still one of the most pervasive means of communication and semiconductor electronics seems to have an important role to play in improving the telephone network and in extending the use that can be made of it as a powerful means of communication.

The telephone system can provide a personal link between every home and, perhaps eventually, the basic structure for linking any two or more individuals wherever they may be. But the system need not be restricted to voice communication: it can also be used to send data in the form of pulses. Such data transfer is obviously of use in commerce, industry and all forms of administration, but could also be employed to the benefit of the individual. Data flowing into and out of an individual's home, perhaps presented on a television screen and organised by a small computer terminal, would radically alter the average way of life. There would be much less incentive to go out to work when contact and information were so readily available at home. Even shopping could be done more efficiently if full knowledge of local choice and price were available in the home. Nor would service visits to homes always be necessary if meters were read automatically and if faults in domestic electrical equipment could be diagnosed from afar. Much schoolwork might be carried on more efficiently from home, the sending of items by post might often be avoided, advertising would alter its approach and votes could be cast from home.[25] The possible changes that could be brought about by a new use of the telephone system by the incorporation of semiconductor electronics are momentous and it is, therefore, worthwhile considering some of the difficulties that have arisen and seem likely to continue to arise.

It may seem ironic that semiconductor electronics, primarily invented for use in telephone switching and associated functions, penetrated much more rapidly into other areas of electronics than into telephony. Certainly the actual switching function, the connection of one line to another, is still usually performed by a mechanical switch or relay, and relays are likely to continue in use for a very long time to come. The semiconductor has now made a considerable impact on the various control functions in the telephone system and developments in this area are rapid. But even this impact was slow to start, mainly because telephone networks are very large and cannot be replaced all at once, and any partial replacement causes great problems of compatibility.

There is a natural and proper reluctance to innovate prematurely in major capital equipment because of the danger of investing heavily into a technology before it has reached maturity. Too rapid innovation is as wasteful, or possibly more wasteful, as lack of innovation. In major investments like telephone exchanges the risk of rapid obsolescence must be avoided.

In the early days of the telephone the basic switching function was achieved by hand. Later, various electro-mechanical devices were introduced and many of these are still in service. Switching can now be achieved by the use of reed-relays controlled by semiconductor electronics. This is the first step towards electronic switching. When we talk about an electronic exchange, we often mean one that uses electronic routing but reed-relay connections, although fully electronic exchanges do now exist. The electronic exchange is faster, takes up less space and is, at least potentially, more reliable and cheaper to maintain,[26] but it has some disadvantages in addition to high capital cost. A serious problem is increased power consumption, because electronic exchanges use power continuously while electro-mechanical devices use power intermittently.

It is estimated that half of all telephone calls in the United States will go through a total of some 4700 electronic exchanges by 1985.[27] In Britain, a programme for building initially eighteen large electronic exchanges in addition to hundreds of small exchanges, was begun in 1971. It is estimated that this British programme will cost some £100 million over a seven year period and that it would cost some £2000 million to re-equip all telephone exchanges.[28] The sheer cost of electronics in switching obviously gives cause for concern, especially as the provision of an adequate telephone network is seen as a public service in most countries. If existing electro-mechanical exchanges are adequate and cheap, there is no obvious mandate to replace them while they still function with systems which, though they are much more versatile, are also much more expensive. As with the electronic watch, semiconductors in switching must compete with an established product, but not in sufficient quantity to allow the sort of mass production which can decimate semiconductor prices. It is far from clear whether a public service should charge the public more for a better service when the alternative is to maintain a cheaper and perhaps perfectly satisfactory existing service.

> Our sole purpose is to provide efficient communication at the lowest
> possible cost.   Dieter Alsberg, Bell Laboratories

The electronic computer is, of course, an established fact now, whereas the fully electronic public telephone exchange is still very much a research proposition. The reason for this is that the electronic computer was developed to do something which really could not be done by any other method, whereas an electronic telephone exchange is in competition with existing electro-mechanical systems which, although they have defects, are capable of doing the job, and so an electronic telephone exchange would not be acceptable unless it did a better job at the same cost. It wouldn't be acceptable if it did a better job at a greater cost.   J. E. Flood

Some indication of the sort of problems involved in trying to use semiconductor electronics in telephone exchanges is provided by the British experience. In Britain, as in many countries, the Post Office has a monopoly in the field of providing a telecommunications service and the maintenance of this monopoly is seen as essential for the provision of an efficient service.[29] One inevitable result is that the sort of accusations that are normally levelled against any giant corporation – particularly that it is bureaucratic and that its decision-making is remote and often incomprehensible – are frequently directed at the Post Office.[30] However the decision-making process of the Post Office on British telecommunications systems is influenced by the telecommunications industry, the government, and others.

In the fifties, most telephone exchanges used the Strowger switching system, dependent on revolving wiper blades to make connections by means of an electro-mechanical, step-by-step process. The Strowger system, like much good modern technology, is based on an idea from the last century. In the thirties, a new electro-mechanical system, known as crossbar, was developed in which contact was made by means of relay springs. Crossbar was improved in the forties and was being rapidly adopted in many countries by the fifties. Yet this did not happen in Britain. The transistor had by then been invented and it seemed to many in the Post Office and in the British telecommunications industry that semiconductor electronics would soon be capable of providing a new generation of electronic, rather than electro-mechanical, telephone exchanges. It seemed that it would hardly be worthwhile introducing an intermediate generation of crossbar equipment. Consequently, while much of the rest of the world, including the United States, changed to crossbar, Britain made successful improvements to Strowger. British manufacturers of crossbar found no market for their product at home and overseas were rapidly overtaken by the Swedes.[31]

> There is a past history of the Post Office making some bad mistakes. Some people find it hard to visualise that it might ever do anything right. J. E. Flood

In 1958, the first fully electronic telephone exchange in Britain was opened as an experiment at Highgate Wood in London. The experiment demonstrated both that electronic exchanges worked and that, because of the large amount of apparatus which had to be attached to each subscriber's line, they were rather uneconomic. Electronic equipment worked with low currents, not the high currents needed to ring the subscriber's telephone bell, feed the microphone and operate the meter. Consequently, it was decided to leave that part of the system electro-mechanical, using reed-relays, and to make only the control function at the central exchange fully electronic. The first such exchange, called TXE1, opened at Leighton Buzzard in 1966 and hundreds of exchanges based on a parallel development for small exchanges, known as TXE2, have been brought into operation since.[32] A new generation of electronic systems for larger exchanges, TXE3, was not accepted, partly because it was argued that newer systems were in the laboratory and would soon be available, but mainly because it was too expensive. By 1967, Britain had a modern electronic system for small exchanges and nothing for large ones. It was then announced that crossbar – neglected for two decades – would be adopted as an interim, emergency measure to replace big city exchanges, most of which dated from the thirties and were urgently in need of replacement. By 1975, 90 per cent of telephone exchanges still worked on the old Strowger system. Not until 1971 did TXE4, an electronic system for large exchanges, begin to be manufactured. TXE4 has caused considerable ill-feeling among British firms newly geared to produce crossbar and there have been accusations that TXE4 is unsuitable not only because it is American, but also because it is by now obsolete; though many would refute both allegations.[33] Any change to a radically new technology is likely to cause severe dislocation in industry, closure of factories, and redeployment of labour. It would appear that British telecommunications provides a fine example of the difficulties experienced all over the world of adapting to the pace of change in semiconductor development.

Relations between the Post Office and its manufacturers seem to have been troubled for some time. In the past, manufacturers approached the Post Office as a consortium, but by 1969, political arguments that this stifled competition had caused government to force the breaking up of this system. The change was unpopular with both the Post Office and the manufacturers and did not improve relations between them. In the United States, A.T. & T. dominates American telecommunications and sees its participation in manufacturing, through Western Electric, as essential to its progress.[34] In Britain, there are no such links.

> If you look at the Bell organisation in America, you find a unity in administration, research and development, operations and manufacture. Having got such an organisation, you could not imagine them splitting the large resources required for switching development into several different organisations to produce competing designs – which has been the equivalent situation in the U.K.
>
> John Martin, British Post Office

In both the United States and Britain, semiconductor manufacturers accuse those in telecommunications of being far too concerned with reliability. Semiconductor devices are inherently more reliable than any alternative, but while the failure of electromechanical devices tends to produce merely poorer service, failure of centralised controls dependent on electronic devices in some systems can produce total shut-down. British system designs aim to reduce this problem by distributing the control, but still find it necessary to ensure that semiconductor devices are thoroughly tested for reliability; some would even say more thoroughly than their predecessors ever were because they are meant to be more reliable and to last longer.[35,36] It has even been said that because some integrated circuits are so utterly reliable, it is hard to measure how unreliable they are.[37]

> Until a process has been around for some years, the Post Office, and I think probably quite rightly, is not satisfied that it has been established as a reliable process and they are quite prepared to sacrifice performance and price in order to get the assurances on reliability. I don't think that the components that you would get in a hand calculator would be of sufficient reliability to satisfy the British Post Office.
>
> John Cave, Plessey

> The ordinary components are reliable enough for them ... The equipment makers tell me that the British Post Office is the most difficult in the world.
>
> British electronics executive

One basic problem is, of course, that any telecommunications system requires massive capital expenditure and it can be no light decision to replace perfectly functional equipment with an expensive, though superior, alternative. There is a problem too in the pace of development. In the semiconductor industry, the pace of change is rapid. A new product one year may be obsolete the next. In telecommunications the new product has to last twenty years – perhaps longer – and to be compatible not only with what is already there, but with what might be introduced in the future. With such problems, it seems essential that those who make decisions about telecommunications retain the closest liaison with the electronics industry.[38] This seems to happen to some degree in the United States, but in Britain, the links between manufacturer and operator are much more tenuous. There is not much interchange of staff between Post Office and industry: most people begin and end their careers in one of the sectors. It seems unavoidable that telecommunications will have some sort of centralised control and direction, but the welcome that has been given in the United States to the Carterphone decision, approving peripheral additions by others to the telephone system, suggests that some would welcome the opportunity to reduce the total control of a central organisation.[39] In Britain, the total monopoly of the Post Office over all forms of telecommunication services means that decisions about future telecommunications are likely to be made by the central authority, albeit influenced by outside needs and pressures.

> If our manufacturers keep on pestering us with new ideas, if they are sound they will get into the system ... We, generally speaking, let them come to us ... We welcome signs of change and practice with them, provided they tell us in advance and we can assure ourselves that they are not going to rush us into headlong disaster.
>
> British Post Office official

The British Post Office seems to recognise that its role is not that of an innovation leader in semiconductor technology.[40] There are advantages in this because much of the risk and expense of innovation will be borne by others, but the hidden costs of an increasingly obsolescent telecommunications system may be considerable and a balance between innovation and caution must be struck. Much of the problem seems to centre on electronic exchanges. Telecommunications systems do have to operate continuously in unfavourable environments and without failure. Such conditions are much more rigorous than those encountered by pocket calculators, electronic watches or even by large computers. Yet, the actual changes that semiconductor electronics has made possible in other sectors and the potential changes such electronics holds for telecommunications mean that hard decisions must be made. Perhaps it would be more realistic to view some of the potential benefits from the use of semiconductor electronics in telecommunications, such as data transmission, as new products, distinct from the accepted telephone service. Rapid progress in telecommunications now seems dependent on acceptance of technological obsolescence rather than merely physical obsolescence.[41] This, with its implications of wasted resources, may very well be thought undesirable not only by the Post Office, but by society as a whole.

The problems which confront those who have to make decisions about the use of semiconductor electronics in telecommunications are massive and almost certainly greater than in any other application. If errors have been made in British telecommunications, they should be regarded much more as indications of the severity of a continuing problem than as proof of long-standing ineptitude. Concern about reliability, compatibility and the rate of development and sheer cost of the new technology can also be seen as sober reservations about the impact this new technology may have and a consequent reluctance to make irreversible decisions. In no field can decision-making be more difficult than in the application of semiconductor electronics to telecommunications.

There are now indications of a new attitude in British telecommunications, more committed to electronics. Such projects as View-data may herald the new era. This is a system in which a central computer is linked to television sets in the home by means of the normal telephone lines and exchanges. Instructions can be sent from the home by means of a small terminal and answers will appear on the television screen. The success of the scheme depends on what demand there is for data in this form and on how cheap the service and the necessary equipment can be made. If the scheme is successful, a full national service should be in operation by 1978 because the service does not require the introduction of a new switching and transmission network.[42] The old problems of semiconductors in telecommunications – those of reliability, compatibility and expense – will still remain, but they will be seen in the perspective of the entirely new sort of telecommunications service semiconductor electronics can offer.

> You may find that you are constructing a society where people, after spending centuries becoming more and more mobile, are now becoming less and less mobile. Why? Because it costs too much in terms of time and resources to be mobile.   Gunther Rudenberg

Many observers both within and without the semiconductor industry see the automobile as one of the greatest potential markets for semiconductor components, particularly integrated circuits.[43] There is scope for the use of semiconductor electronics in vehicle safety with such devices as anti-skid braking controls, in engine control with electronic ignition and fuel injection, in instrumentation, servicing and probably a great deal more.[44] Yet this scope has been recognised for some time. In 1969 it was hoped that, by 1974, each of the ten million American cars produced annually would contain perhaps twenty integrated circuits. The impact of this market on both the semiconductor and the automobile industries would have been substantial, but for various reasons the market has not yet materialised.[45] There are signs that it may very shortly do so.

One of the main problems is very similar to that experienced in the adoption of semiconductor electronics in telecommunications. While telecommunications networks have massive capital sunk in existing systems, the automobile industry has its capital invested in huge, mechanised production lines. Such a method of production means that even minor change is expensive and can hardly be undertaken as an experiment. Moreover, the automobile industry has to live with any change it makes for a number of years, while change in the semiconductor industry is very much more rapid. Consequently, incorporation of semiconductor devices in automobiles almost inevitably means that the vehicles will very shortly contain obsolescent electronics. There are other problems too. The conditions inside a car engine are rigorous; hotter, noisier and dirtier than conditions experienced in even a missile. While a component in a rocket may have to work for only a few minutes, that in a car must function for years and it must do this for a tiny fraction of the price of a missile component. Moreover, the car must be serviced and repaired by local mechanics, knowledgeable in the problems of moving parts, but totally unfamiliar with semiconductor electronics. While the rapidly declining price of integrated circuits has made their powers of logic increasingly attractive to automobile manufacturers, such price reductions have not been evident in much of the sensor equipment needed to feed information about the car's performance to the integrated circuit.[46]

> Semiconductor technology is far ahead of what the automobile companies themselves are ready to accept and incorporate in their designs
> ... I think, in general, they are somewhat suspect of what integrated circuits are ... They don't have a good feel for what the potential capabilities will be of this particular technology and they are just, in a sense, not ready to accept it to the extent that they could.
>
> Dean Toombs

Yet these logical objections to semiconductor electronics do not meet with heartfelt sympathy from the semiconductor industry. Many there see the main problem as being that of vast and bureaucratic automobile corporations uninterested in major innovations, a view shared by men who have worked in both industries.[47] As one semiconductor expert put it, the basic engine technology of his new car was really old enough to vote.[48] Viewed from the world of semiconductors, this seems so strange that it smacks of negligence or indolence. Such an opinion is the easier to form when it is remembered how vast the market for semiconductors will be when they at last enter the automobile in quantity.

Some semiconductors have been used in cars for a considerable time, but often merely in such gadgetry as automatic windows and stereo equipment. In the United States, safety and health legislation has recently virtually forced the adoption of more electronics. Many new cars appeared with sobriety testers, with engines that would not start until the seat belt was worn and with electronic fuel injection to reduce noxious emission. Such electronics appears to have been unpopular with manufacturers and users alike for the economic depression of the mid seventies has put much of the legislation in abeyance and the electronics has frequently disappeared.[49] One semiconductor expert, particularly keen on the use of semiconductor electronics in automobiles, had personally disconnected the safety electronics in his own vehicle. The automobile manufacturers seem to assume that semiconductor electronics in cars is not the sort of thing that will make their product more attractive and, hence, more competitive. Instead, electronic additions are made as optional extras from which a profit is expected on each item, much as the calculator manufacturer expects a profit on each calculator he sells. In Europe, where car accessories are often added by individual owners, there is now a considerable market for semiconductors in electronic ignition and in such accessories as tachometers. In the United States, the individual does not add accessories. He expects his car to come fully equipped and if the manufacturer is unenthusiastic, then he is even more so. There are few vehicle accessory shops in the United States and those there are generally sell tachometers which operate on the old sticky cable principle.[50]

> It has been very difficult to come up with another hand-held calculator ... There is no readily apparent product like that that is the next candidate that one could essentially execute as an integrated product to open up a new market without engaging somehow or other with other industries such as the automotive industry, such as telecommunications ... but in order to make that happen you must essentially revolutionise that industry.   Dean Toombs

One of the most revolutionary of possible changes would be the use of a microprocessor to monitor and control all the mechanical parts of a car. Work in this direction seems to have been going on for some time with particular interest shown by Chrysler, using a team and facilities once financed by NASA.[51] It seems that the automobile manufacturers also have an interest in automatic braking and blind driving aids.[52] Perhaps a more likely initial move into semiconductor electronics which goes well beyond the gadget stage is the completely electronic instrument panel, giving a great deal more information than conventional displays, more clearly, cheaply, in a more compact and logical form and only when such information is wanted.[53] If this is to be the first step, it will prepare the way for many more major developments as both the automobile industry and the semiconductor industry become committed to basic electronic change in cars.[54]

Perhaps the area in which and through which semiconductor electronics has had the most profound impact is computing. Computers of a sort were available before the invention of the transistor; some rudimentary ancestors of the modern computer are centuries old. What the transistor has done – and integrated circuitry even more so – is permit the construction, at reasonable cost, of machines of immense complexity and capability. The potential utility and growing ubiquity of the computer, rather than just the computer itself, are very largely a consequence of developments in semiconductor electronics.

The benefits that could be gained from a calculating machine have long been evident and accounts of struggles to implement logical functions with primitive machinery make fascinating reading.[55] Electro-mechanical and electronic equipment was an immense improvement, but attempts to build large computers before the transistor was available confronted all the disadvantages presented by the regular replacement of thousands of hot, fragile, power-consuming valves. A very early electro-mechanical computer, built at Harvard University in 1944, contained three-quarters of a million parts, 800 km of wire, was 17 m long and 3 m high.[56] The first fully electronic machine, the ENIAC, built by the University of Pennsylvania in 1946, weighed no less than 30 tonnes, occupied 150 m² of floor space and had 18 000 valves and diodes to be changed frequently.[57] Such monsters were built by university specialists, in both the United States and Britain, and often for highly specialised purposes. By modern standards, their computing power was tiny and their social and economic influence insignificant.

Transistors brought the beginning of a transformation. By the mid fifties, transistors were being made that were sufficiently fast and reliable for computer applications. Hot, unreliable and power-consuming valves were no longer necessary and became obsolete for computers by 1958. By 1964, the transistor too was being phased out of computers with the development and rapid acceptance of integrated circuits. Large companies, often with their roots in mechanical calculators, took over the development of computers from the universities and often still dominate an exploding market. IBM, for example, can trace its ancestry to a company making punch card machines in the 19th century. It now controls 70 per cent of the world computer market.[58] Much of this success can be attributed to the introduction of complete ranges of fully compatible machines in the mid sixties to cater for a much wider variety of users.[59] Much of the present impact of computers is a consequence of this rapid diffusion of computing rather than of simply a growth in the power of computers.

No other application of electronics reflects the developments which have taken place in the semiconductor field over the last twenty years quite as well as the computer. It is not so long since it was imagined that a handful of computers would satisfy even the most generous estimations of need. A decade after the invention of the transistor, it was still hard to see beyond the computer as merely the key to automation.[60] In fact, the computer has become much more than that and promises to become spectacularly more important. Many people feel uncomfortable with even the present impact of computing and foresee real dangers in any growth of this impact. What, then, do computers do and what more are they likely to do?

A computer stores information and is programmed to perform logical functions with that information. The value of a computer lies in the huge amount of information it can store and in the high speed with which it can function. In terms of mass and speed of work, the computer can do more than is humanly convenient and often more than is humanly possible.[61] In the most general terms, it can perform both clerical and analytical tasks, though this tends to underplay both the actual and the potential uses of the computer.[62] Despite the heralding of the early computer as the 'Electronic Brain' and the foreboding with which many people regard the more revolutionary uses of computers, the majority perform the most tedious of tasks. Most computers used in commerce, for example, are engaged in payroll, stock control and invoicing work.[63] The myriad of uses to which computers are put in other fields is too familiar to bear much repetition here. It will be sufficient reminder to note that, without the computer, the present organisation of banking, insurance and much else in the commercial world would be radically different. In industry much machine control and automation of all sorts is dependent on the computer. Without it, space exploits would be impossible. In communications, the computer is used in telephone exchanges, for traffic control and to regulate air travel. In education, it is used for teaching and has become essential for research in many disciplines. The computer is becoming just as important in libraries, for police work, in medicine, in meteorology and in all manner of administration.[64] It is notable that most change is being brought about by small and often very cheap computing elements known as microprocessors rather than by large, expensive computers. The availability of the microprocessor opens up innumerable possibilities for automation and control of processes in industry, commerce and even in the home.

It is precisely because the computer is so universally useful, that there is so much concern about its present and future socio-economic impact. Some see the computer as a cause of unemployment, a threat to privacy and a substantial first step towards a state where decisions are taken by the few and control is exercised by machines.[65] Stories abound of errors made by computers and of infuriating difficulties in getting such mistakes rectified. Less attention is given to the benefits enjoyed because of computers and this is probably because most people feel those benefits only indirectly. It is difficult to quantify the personal advantage of a more efficient police force, an automatic assembly line or a picture from Mars. Future uses of computers will very likely exert a much more direct impact on the individual, and perhaps one for which he is neither prepared nor willing.

Extremely rapid development of computers and of computer usage is expected in the remaining years of this century. A forecast made in 1968 estimated that there would be a 50 per cent reduction in the labour force of existing industries by the late eighties and that computers would be as common in private homes as telephones in the year 2000. By this time, the price of computers will have declined by a factor of 100 and machines will be available, capable of learning from their own experience. The present money and cheque system will vanish. Cars, for example, will have computer autopilots and urban traffic will be completely controlled by computer. Teaching, libraries and medical research and diagnosis will be particularly influenced.[66] Such forecasting is prone to massive error, but it may give some indication of future developments.

Whatever the computer may be used for in the future, it will be dependent, as it is now, on semiconductor electronics. Without the transistor and now the integrated circuit and the semiconductor memory, computers would have remained primitive and their influence would have been too small to have caused much comment or concern. In fact, computers are now so pervasive that they are responsible for the design of the very integrated circuits upon which they are dependent. Thus, not only does much of the future influence of computers depend on developments which may take place in semiconductor electronics, but much of the future of semiconductor electronics depends on what progress is made in computers.

To test foresight with hindsight may not be a rigorous exercise, but nothing connected with foresight is rigorous except this statement itself. If a scenario for 1975 had been written in 1950, it is doubtful whether it could have envisaged the true integrated circuit. It seems that the single chip containing dozens of components would have required a leap of the imagination more appropriate to the science fiction writer than to the technological forecaster. The scenario would, therefore, probably have contained only devices and applications based on transistors and rectifiers. Presumably a healthy development of these devices would have been forecast, so that they would have become cheap, reliable and mass produced. The assumption that the frequency response would be improved, the power handling capacity increased and circuit design improved seems entirely reasonable. Equally reasonable seems the assumption that packing densities and wiring techniques, together with passive components, would be developed to a high pitch of perfection.

When a technology is new, its possibilities are often all too obvious and tend to produce unbounded optimism. As the technology matures, the difficulties it presents – in manufacturing, in its relations with other technologies, in finding the best applications, in marketing and so on – become increasingly apparent and much more sober predictions prevail. The history of semiconductor electronics has provided some illustration of this phenomenon. It has also demonstrated the divergence of opinion that can exist, even among experts, concerning the prospects of various technologies. Considering the speed of semiconductor development and the present massive breadth of semiconductor application, it is hardly surprising that there is still no consensus of opinion about what the future holds. The few examples of semiconductor applications we have given may serve to indicate some of the pits into which the unwary technology forecaster may fall. Semiconductor technology which would radically change the automobile and telecommunications has been available for some time, but, for a variety of reasons, has not yet been fully utilised. On the other hand, the use of semiconductor technology in calculators and watches has had a far greater impact than was generally anticipated. In computing, the integrated circuit was the key which triggered an explosion of activity in the sixties. Predictions were made then which would have seemed irrational or irresponsible a decade before, but which now seem more likely to have been underestimates of the power and influence of computing. With the present development of minicomputers and microprocessors there is available a hierarchy of computing power, from the calculator to mammoth computers, which will be formidable indeed and which will provide the most significant social impact yet of semiconductor electronics.

In previous sections we looked briefly at the forecasts that were

made on various semiconductor applications and saw the great in-
accuracies in all forecasts and predictions. One may ask the purpose
of forecasting, if accuracy is never achieved. The answer to this question can be given at two levels: the human need to look into the
future, for we cannot be satisfied without hope and plans and
speculation. The other level is a practical one. Although we cannot
know the future, we nevertheless create it by all we are doing.
Forecasting then becomes an essential part of an attempt to create
the future we desire rather than drift into an undesirable one. To
shape the future we want, we must attempt to understand the
relationship between our present endeavours and their future implications. It is a basic dilemma of the human condition that we cannot know the future yet must know how to shape it. In technology,
these problems have recently acquired an acuteness they did not
have in the past. Not only has technology become all pervasive but
some decisions have become extraordinarily fateful. Technology
determines the wealth of nations, technology determines questions
of war and peace, technology shapes our environment and
technology determines whether we have a civilised future. We must
make technological choices, and we must make them with as much
informed concern for the future as is possible. 'For generations it
has been taken for granted that all that *can* be done in science
and technology *must* be done. The new ethic emerging ... is
that somehow man must agree not to do all he is capable of
doing.'[67]

In the full knowledge that our chances of being even partially
right are slim, and our chances of being totally right negligible, we
shall indulge in a little crystal ball gazing into the future of semiconductor electronics.

No doubt, the present state of development of solid state electronics and of computers will lead to more automation. What is
much harder to tell is what form this automation will take and what
social repercussions it will have. In the factory, automation may
change in quality and machines which may legitimately be called
Robots may take over many human tasks. Robots can assemble cars,
they can weld, operate warehouses, spray paint. Remotely controlled Robots may mine coal or operate chemical plant. The
problems associated with these possibilities are great and most of
them are not technical. Will Robots create enough wealth to give
employment to the many workers they are bound to displace? Can

people to be engaged in high technology maintenance and control work be given the necessary education and training and will they achieve the social status they require? Will human production line workers tolerate sharing work with Robots, who make no demands, take no breaks, drink no tea and provide no companionship? It may be possible to do much of the work now done by divers without human presence underwater, and nuclear operations, should these continue to be used, will become more automated than at present. Large-scale automation may cause a number of present problems to accelerate and reach alarming proportions. The greatest incentive for automation exists in countries with high labour costs, and this means the most advanced and richest countries. By introducing automation selectively in this way, their economic superiority will be further strengthened and the gap between rich and poor nations will grow. Problems of exhaustion of natural resources will also become more severe, for increased productive capacity calls for increased production. In some way this fact may alleviate the rich nation – poor nation problem, for the owners of natural resources often are relatively poor and will be able to exert pressure on the rich nations to pay higher prices for their raw materials. A complex situation fraught with dangers and difficulties. In the rich countries themselves there may be problems of endemic unemployment and problems of too much leisure. Perhaps people need to spend a reasonable part of their life working for remuneration, as without this obvious valuation of their activity they may find it difficult to give value to their own efforts. Not many people can be happy without some external endorsement of the value of what they are doing. Non-productive work, cultural work, organised games of various kinds may become essential substitutes for the work taken over by machines.

Further automation will not only affect the factory and its products, it will also affect administrative work of all kinds. Already stock-keeping, payrolls, accounting, invoicing and similar tasks are being done by computers. Gradually filing is becoming computerised. It may soon be possible to have routine letters typed by computer, mail sorted and routed by computer and office workers working from home and joined by various data links. These developments will not, we suspect, go as far as would be technically feasible, simply because office workers at home would become too frustrated with loneliness. The companionship of the office is important to the well-being of the organisation and of the workers themselves.

Shopping will probably continue in its move to automation. There seems no reason why warehousing should not be completely automated. Goods could be moved into and out of the warehouse without human interference; computer controlled handling devices, Robots, could do it all. This would be particularly advantageous with frozen goods, where working conditions are extremely unpleasant for humans. The retail shop could also be fully automated. A customer would inspect the goods, which he must do in order to make his choice. He would press the appropriate button for the goods he wants and these would await him, ready packed, at the check-out point. To obtain them, he would insert his credit card and the goods would be moved to his vehicle or to a delivery van. The computer would debit the customer's account, see to the accounts of the shop, to its tax liability, and also re-order the goods sold. The order would be suitably accumulated and dispatched, still by computer, to the appropriate warehouse. Labour in retailing would be minimal; a few managers, computer operators, display arrangers, maintenance workers, supervisors and drivers. Severe social problems can again be foreseen. Retailing now provides much work opportunity for the less highly skilled. Their jobs would largely disappear and it is difficult to see what would replace them. It is also possible to remove the visit to the shop altogether and replace it by ordering goods from home. These might be displayed on a television screen and could be ordered by a data link from the home. This may sound very convenient and would certainly make sure that price competition cannot be avoided by the shopper's inability to 'shop around', but it would further increase the already severe isolation of many people and remove one of their incentives for venturing outside their homes.

The home itself may well become the scene of much automation. Already electronics is finding a considerable outlet in the home, but so far mainly in home entertainment: television, radio, record players, tape recorders. There undoubtedly is some scope for further development in this area, for instance miniature television sets, but the main developments will occur in areas in which electronics has as yet made little progress. All our domestic machinery, cookers, heating appliances, washing machines, refrigerators, freezers, dish-washers, etc., are almost entirely devoid of electronics and are controlled by crude mechanical means. With the availability of solid state electronics capable of handling and controlling large currents at very reasonable prices, there is tremendous scope. Intricate controls for washing machines, cookers completely programmed and timed either remotely or as much in advance as desired (perhaps a weekend ahead), automatic vacuum cleaners, automatically controlled lights, electronic locks, burglar-proofing, electronic games, automatic air conditioning; these and many more are very real possibilities for electronics in the home of the future. Many of these features might be controlled by a central computer in the home. Perhaps obsolescence will not be built so much into the machines as into their controls, and this might be a good thing from the point of view of conservation of natural resources.

The household will be considerably affected by any developments in telecommunications. Using a modified television set, access can be gained to all kinds of information and all manner of transactions may be carried out. The newspapers may largely be displaced by journal pages displayed on the TV screen and these can be continuously updated. The effect on journalism is bound to be drastic and not necessarily beneficial; the more ephemeral the word becomes, the less care will be put into producing it. 'Words writ in water' will become an electronic reality. Other display facilities may become available on the domestic television screen: programmes of theatres and other forms of entertainment, together with a selection of critical reviews and the possibility of purchasing tickets; stock exchange information and automatic access to the stockbroker or his computerised equivalent. It may even be possible to dial a word and obtain the relevant entry from the *Encyclopaedia Britannica* or other reference works. Many more services can be envisaged: baby surveillance cameras, bedside stories piped directly to the cot, airline reservations from the home, holiday displays and bookings. Perhaps the provision of some of these services will require so much effort as to provide work for all the people the services will displace. Certainly the acquisition of all these conveniences will require a lot of hard and boring work.

An area in which electronics is likely to move quite rapidly is that of monitoring of body functions. A device to give warning of a malfunction of the heart, enabling the prevention of an acute coronary attack, has already been devised. Heart pacemakers are now in wide use and enable people, who without them would be severely disabled, to lead a normal life. Altogether the scope of electronics in preventive medicine, diagnostic medicine and perhaps in the treatment of disease has enormous potential. It may be possible to have completely automated routine screening with a diagnostic computer print-out. The main problems that remain to be solved are probably the knowledge of what parameters to measure and the development of suitable transducers to translate the parameter into an electrical signal. It may be possible to monitor a function and automatically release the required amount of a drug to counteract any malfunction. In this way the control of diabetes, to mention but one example, may be completely automated. This would overcome many difficulties of present day treatment, because drug administration is notoriously inaccurate and difficult to adapt to varying requirements. There is also vast scope in aids for invalids. Any device designed to enable Robots to fulfil tasks could be adapted to help invalids: radar for the blind, radar controlled carriages for them, intricate manipulators for the paralysed, improved hearing aids, and many more devices to give independence and enjoyment to those most in need of it. Shall society be caring enough to provide all these things?

The list of future applications could be extended; the limit is only lack of imagination and fear of boring the reader. But before we leave the future entirely, we must look at some of the potentially sinister aspects of the use of solid state electronics. Bugging and surveillance devices are easy to produce and easier to misuse. Computers have an unlimited memory with unbelievably fast access and this can be used to store socially useful information, for example to keep criminal records. It can also be used to file information on citizens, on their views, their associates, their bank balances, their habits. Such information can be grossly misused, especially in an unfree society. But even a free society must guard against the possibility of criminal misuse or of simple abuse by bureaucratic organisations.

From the brief excursion into the future, we return to the present era of a society struggling to control its electronic capability. The list of present applications of solid state electronics is formidable, and to make it exhaustive would require a computer print-out. Only electronics can do justice to electronics. Here are a few applications which are illustrative of the trends and of the pervasive nature of electronics. Above all, they illustrate some of the problems that may be caused and must be controlled.

The first set of problems may be called uniformity and depersonalisation. The very large scale of production possible by automation squeezes out of the market anything that is not massively popular. The scope for being idiosyncratically different is given only to the very rich. The computerisation of invoices, personal records, membership of organisations, etc., removes the personal touch and makes it extraordinarily difficult to rectify any errors. Who has not written to an organisation in an attempt to redress a wrong, only to be confronted by a repeat print-out of the original offending message? A computer does not make allowances and cannot tolerate small anomalies; the holes are round and square pegs cannot be accommodated. The technological trend up to now, which was for large computers, encouraged the growth of massive organisations. Now that the minicomputer is arriving, this trend to centralisation may or may not be reversed. On past experience it ought to be reversed, although it is likely that the centralisation versus decentralisation issue is destined to swing backwards and forwards for ever, without ever reaching a stable equilibrium.

Computers are gaining an increasing reputation in decision-making. They are gradually usurping the role of the expert, for if a decision is supported by computer analysis of available options and their consequences, it achieves an aura of infallibility. Unfortunately, computers are rather like sewers; what you get out of them depends on what you put into them. The respectability of a computer-aided decision can be spurious.

The last few decades have seen a steady shift from productive labour, in the sense of labour engaged in the manufacture of goods, to service labour. Most of the services are not, however, repair or maintenance services. In fact their decline is one of the penalties of mass production. The services are in administration, which has achieved massive proportions, in health, transport, education, tourism, etc. No doubt the advent of semiconductor electronics has accelerated the process and caused it to go further than it might have gone. There are signs that these processes have had and will have serious social repercussions at all levels. The traditional working class has decreased in size and this is causing changes in the political structure and some strain on family loyalties. The lack of employment opportunities in production jobs has caused unemployment which may well be structural rather than temporary. The types of employment which have vanished are those most suitable for people with middling to poor intellectual capacity, and these people find it increasingly difficult to obtain satisfactory employment. The jobs left are often the really badly paid ones, such as in catering, or the really unpleasant ones, such as refuse disposal. There are three conceivable remedies to this situation. We must have continuing growth in production, but we must select our technologies judiciously if we are to avoid an unsupportable burden on the natural resources of the planet. Alternatively, we might reinstate craftsmanship and increase services, but this can be done only if we have the wealth and the political will to do it. The third possibility is to reduce the time spent at work. To do this, we again require the political will to share out the work and its fruits and to cope with the provision of leisure activities on a large scale. The political will is the dominant issue, for the level of wealth adequate to do these things is a political decision.

The growth in communications caused by increased availability of solid state electronics has given a tremendous fillip to our awareness of the smallness of Earth. Whether we consider the instant availability via satellite of news coverage from all parts of the globe, or the ease of booking airline flights by a computer booking system, or the space exploration made possible by modern electronics; all these things have shrunk the size of our planet. We have become more aware of other people and we have come to recognise the finiteness of our resources. We are too close to these developments to appreciate the nature of their impact on our collective consciousness and our political institutions, but in the fullness of time the influence is bound to be great.

There can be no doubt that the Military played an important role in the development of semiconductor electronics. Whether one takes the view that the role was decisive or not, no one can doubt that it was substantial. And in military terms the effort has paid off handsomely. Virtually the whole arsenal of weapons in a modern army, navy or air force is packed with electronics from top to bottom. Ballistic missiles, cruise missiles, radar, sonar, navigational systems, homing missiles, spy-satellites and all the rest of the awesome list, they all depend in vital aspects on solid state electronics. But not only the weapons, even the military administration system is now as heavily computer dependent as any administration of a large organisation. Even military decisions can be computer aided and one speaks of the automated battlefield. Test ban treaties are dependent on electronic detection methods and may increase our security, but such renegade feats of electronic technology as cruise missiles may decrease the stability of the balance of terror. There can be no doubt that armed forces the world over are using modern electronics, though whether any citizen of any country is more secure for their use is a debatable issue.

The quality and reliability of modern electronics is almost unbelievably high. An integrated circuit is a truly awe inspiring achievement of human inventiveness. It performs the most intricate functions with complete dependability at extremely low cost. The limiting factor on performance is now usually to be found in the peripherals. A pocket calculator is only as good as its keys, battery connections and case. The electronics hardly ever fails, provided it is properly treated. In the same way, an electronic watch is only as good as its display, its case and its interconnections. The semiconductor industry grew up with extraordinary precision and attention to detail built into it right from the start. Other industries, brought up in the more robust traditions of the mechanical age, find it difficult to match the perfection of electronics. Perhaps this is one of the causes of the long delay in introducing electronics into the motor car. Similar difficulties may arise when electronics begins to penetrate the domestic appliance and like markets. But no doubt these difficulties will be overcome and electronics will continue in its penetration of all aspects of technology.

Any history of an innovation distorts the truth in one particular way: it describes the successful innovation and disregards the many unsuccessful ones. This is inevitable, because only successful innovations reach the light of day sufficiently to draw attention to themselves and only successful innovations can be traced from documentary sources. But for every successful innovation there must be many failures, for innovation has an element of gamble in it. Inventions are like lottery tickets thrown into a hat, with successful innovations as the winning tickets. And the larger the hat the more winning tickets there are in it. All other things being equal, the number of successful innovations depends on the number attempted. Other things are not, of course, always equal, but nevertheless it seems inevitable that there should be many failures accompanying every success. The failures sink without trace and the historian has no means of telling their tale. They represent a lot of wasted resources, of disappointed hopes, of misspent energies. It is one of the purposes of any study of innovation to try and learn from success in order to give attempts a better chance and waste fewer resources, not the least of which are hope and enthusiasm.

It is possible to point out a few semiconductor innovations which looked promising at some stage and have yet to hit the jackpot. There was much talk of electroluminescence as the light source of the future, yet it still occupies a very modest niche in modern technology. The picture-phone has yet to prove a commercial success and even the much praised laser does not find many uses outside the scientific research laboratory. The video recorder has been available for many years, but has yet failed to sell in any quantity to the home user. It is less easy to point to actual semiconductor electronic devices which did not make it, but no doubt there are many and we have seen some past examples.

The benefits obtained from a technology cannot be turned into numbers to draw up a balance sheet. Some benefits and damage are caused indirectly and, in any case, any judgement of benefits is value-laden and highly subjective. It would be nice to draw up a cost-benefit analysis and say, 'Humanity has benefited by a net $X million from semiconductor devices'. It cannot be done; all we can do is use coarse overall criteria. Does the technology drain scarce natural resources; does it use a lot of scarce energy; does it produce desirable or undesirable social consequences; does it produce high quality goods? Applying these sort of criteria to semiconductor electronics, our verdict might read something like the following.

Despite some misgivings about the social effects of semiconductor electronics, it seems that on the whole it has been either a useful or at least a neutral technology. Silicon is one of the most abundant elements in the earth's crust, so there is no direct problem of exhausting natural resources. A good deal of energy is consumed in the manufacture of electronic devices, but it is doubtful whether this is a serious contributor to our overall energy consumption. Certainly their operation requires very little energy. Semiconductor technology imposes no limits on its growth; this is the kind of economic growth that is tolerable because it causes no environmental problems. Electronics can be employed in any technology, large or small scale. The flexibility of electronics is such that it can be used in almost any task. The critical choices concern the adaptation of technology to our needs. We have irrevocably entered the electronic age and it is up to us to make it a good age in which to live.

References

Chapter 1

1 C. Freeman, *The Economics of Industrial Innovation*, Penguin, London (1974)
2 F. R. Jevons, *Science Observed*, Allen & Unwin, London (1973) ch. 6
3 Interview with Drs Haayman, van Santen, van Vessen and Mr Tummers
4 J. Langrish, M. Gibbons, W. G. Evans and F. R. Jevons, *Wealth from Knowledge*, Macmillan, London (1972)
5 C. W. Sherwin *et al.*, *First Interim Report on Project Hindsight*, Office of the Director of Defense Research and Engineering, Washington D.C. (1966)
6 R. R. Nelson, 'The link between science and invention: the case of the transistor', in *The Rate and Direction of Inventive Activity – Economic and Social Factors* (U.S. Bureau of Economic Research), Princeton University Press, Princeton (1962) pp. 549–83
7 C. Freeman, *The Economics of Industrial Innovation*
8 E. Mansfield, *The Economics of Technological Change*, Longman, London (1969)
9 *TRACES (Technology in Retrospect and Critical Events in Science)*, National Science Foundation, vol. 1 (1968)

Chapter 2

1 E.g. *Dictionary of Scientific Biography*, American Council of Learned Societies, Charles Scribner's Sons, New York, vol. 4
2 *Dictionary of Scientific Biography*, vol. 5 and interview with G. v. Minnigerode and H. Grundig
3 E.g. *Dictionary of Scientific Biography*, vol. 9
4 E.g. E. Grimsehl, *A Textbook of Physics*, Blackie & Son, London, vol. 3 (1933) p. 612
5 E.g. T. L. Martin, *Electronic Circuits*, Prentice-Hall, New Jersey (1955) ch. 15
6 J. A. Fleming, *The Principles of Electric Wave Telegraphy*, Longmans, Green and Co., London (1908) ch. VI
7 W. H. Eccles, *Wireless Telegraphy and Telephony*, Benn Brothers, London (1918) p. 281
8 *Nobel Lectures, Physics, 1901–1921*, Elsevier, Amsterdam (1967) pp. 226–47

9 G. W. A. Dummer, 'Electronic components – past, present and future', *Electronic Components*, Special Supplement (9 October 1970)

10 J. N. Shive, *The Properties, Physics and Design of Semiconductor Devices*, Van Nostrand, Princeton (1959)

11 Interview with W. G. S. Parker

12 K. J. Dean and G. White, 'The semiconductor story', *Wireless World*, **79**, 1447 (1973) 2

13 R. W. Pohl, *Otto von Guericke als Physiker,* Deutsches Museum, Abhandlungen und Berichte, Berlin (1936)

14 J. T. MacGregor-Morris, *The Inventor of the Valve, A Biography of Sir Ambrose Fleming,* The Television Society, London (1954)

15 J. Scott-Taggart, *Thermionic Tubes,* The Wireless Press, London (1921) ch. 2

16 R. L. Petritz, 'Contributions of materials technology to semiconductor devices', *Proceedings of IRE,* **50** (1962) 1025

17 G. L. Pearson and W. H. Brattain, 'History of semiconductor research', *Proceedings of IRE,* **43** (1955) 1794

18 *Dictionary of Scientific Biography,* vol. 6

19 W. Wilson, *A Hundred Years of Physics*, Duckworth, London (1950) ch. XIII

20 G. Gamow, *Biography of Physics*, Hutchinson, London (1962) pp. 225–34

21 Interview with R. W. Pohl

22 *Ibid.*

23 R. Hilsch and R. W. Pohl, 'Zur Photochemie der Alkali – und Silberhalogenidkristalle', *Z. Phys.*, **64** (1930) 606

24 G. Glaser and W. Lehfeldt, 'Der lichtelektische Primärstrom in Alkalihalogenidkristallen in Abhängigkeit von der Temperatur und von der Konzentration der Farbzentren', *Nachr. d. Gesellschaft d. Wissensch.*, Göttingen, 2 (1936) 91–108

25 N. Mott, 'Atoms in contact, components of the solid state', *New Scientist* (25 March 1975) 663–6

26 B. Gudden and R. W. Pohl, 'Über lichtelektrische Leitung im Selen', *Z. Phys.*, **35** (1925) 243

27 R. W. Pohl, *Mitteilungen des Universitätsbundes Göttigen*, vol. 15 (1934)

28 R. Hilsch and R. W. Pohl, 'Steuerung von Elektronenströmen mit einem Dreielektrodenkristall und ein Modell einer Sperrschicht', *Z. Phys.*, **3** (1938) 406–7

29 W. Wilson, *A Hundred Years of Physics,* ch. XVIII

30 A. Wilson, 'Theory and experiment in solid state physics', *Bulletin Inst. Phys.* (July 1963) 173–80

31 William Shockley, *Electrons and Holes in Semiconductors*, Van Nostrand, New York (1950); J. N. Shive, *The Properties, Physics and Design of Semiconductor Devices*

32 E. W. Herold, 'Semiconductors and the transistor', *J. Franklin Inst.*, **259** (February 1955) 89–106

33 H. K. Henisch, *Metal Rectifiers,* Oxford University Press (1949)

34 C. A. Hogarth, 'The transistor – its invention and its current prospects',

Physics in Technology, **4,** 3 (1973) 173–86

35 V. E. Bottom, 'Invention of the solid state amplifier', *Physics Today* (February 1964) 246

36 J. B. Johnson, 'More on the solid state amplifier and Dr Lilienfeld', *Physics Today* (May 1964) 60–2

37 'Obituary, Julius E. Lilienfeld', *Physics Today* (November 1963) 104

38 T. L. Thomas, 'The twenty lost years of solid-state physics', *Analog Science Fact, Science Fiction* (1962) pp. 8–13, 81

39 H. Weiss, 'Steuerung von Elektronenströmen im Festkörper', *Physikalische Blätter* (1975) pp. 156–64

40 W. Gosling, 'The pre-history of the transistor', *Radio and Electronic Engineer,* **43** (1973) 10

Chapter 3

1 S. Zuckerman, *Scientists and War,* Hamish Hamilton, London (1966)

2 R. L. Petritz, 'Contributions of materials technology to semiconductor devices', *Proceedings of IRE,* **50** (1962) 1025–38

3 K. J. Dean and G. White, 'The semiconductor story', *Wireless World,* **79** (January 1973) 2

4 Interview with Frederick Seitz

5 V. A. Johnson, *Karl Lark-Horovitz,* Pergamon, Oxford (1969)

Chapter 4

1 Charles Weiner, 'How the transistor emerged', *IEEE Spectrum* (January 1973) 24

2 William Shockley, *The Invention of the Transistor – An Example of Creative-Failure Methodology,* National Bureau of Standards Special Publication No. 388 (May 1974) p. 74; 'The path to the conception of the junction transitor', *IEEE Transactions on Electron Devices,* **ED-23,** 7 (July 1976) 611

3 E.g. Mervin Kelly, 'The Bell Telephone Laboratories – an example of an institute of creative technology', *Proc. of Royal Society,* **203A** (1950) 287–301

4 *Ibid.,* p. 288

5 Ray Connolly, 'Does Bell Labs live up to reputation?', *Electronics* (26 December 1974) 51–2

6 Prescott Mabon, *Mission Communications. The Story of Bell Laboratories,* Bell Laboratories, New Jersey (1975) p. 164

7 Ray Connolly, 'Does Bell Labs live up to reputation?', pp. 51–2

8 M. D. Fagen (ed.) *Impact,* Bell Laboratories, New Jersey (1972)

9 Francis Bello, 'The world's greatest industrial laboratory', *Fortune,* **58** (November 1958) 214

10 'The silliest anti trust suit of all', *Fortune,* **91,** 1 (January 1975) 62

11 'The new frontier', *Bell Telephone Magazine,* **32** (1953)

12 Mervin Kelly, 'The Bell Telephone Laboratories'

13 E.g. Francis Bello, 'The world's greatest industrial laboratory', pp. 148–57, 208–24

14 Ralph Bown's Foreword in William Shockley, *Electrons and Holes in Semiconductors*, Van Nostrand, New York (1950) p. vii; Ralph Bown, 'The transistor as an industrial research episode', *Scientific Monthly*, **80** (January 1955) 45

15 Mervin Kelly, 'The Bell Telephone Laboratories', pp. 148–57, 208–24

16 J. A. Morton, *Organizing for Innovation,* McGraw-Hill, New York (1971) p. 124; H. W. Bode, *Synergy: Technical Integration and Technological Innovation in the Bell System,* Bell Laboratories, New Jersey (1971) p.19

17 Interview with Alan Goss

18 Interview with Dieter Alsberg

19 John Jewkes *et al.*, *The Sources of Invention,* Macmillan, London (1969) p. 215

20 Arthur D. Little Inc., H. G. Rudenberg, *Management Factors Affecting Research and Exploratory Development* (contract No. SD. 235) (April 1965) pp. 103–6

21 E.g. Ray Connolly, 'Does Bell Labs live up to reputation?' pp. 51–2

22 Interview with Walter Brattain

23 *Ibid.*

24 William Shockley, 'The invention of the transistor: an example of creative-failure methodology', *Solid State Devices,* Institute of Physics, London (1972) p. 56

25 Interview with Frederick Seitz

26 *Ibid.*

27 William Shockley, 'Creative failure methodology', *Electronics and Power* (22 February 1973) 59

28 Shirley Thomas, *Men of Space,* Chilton Publishing, Philadelphia (1962) p. 178

29 William Shockley, 'Creative failure methodology' (1973) p. 59

30 William Shockley, 'The invention of the transistor' (1972) pp. 57–8

31 'Walter Brattain Autobiography', *Bell Laboratories Record* (December 1972) William Shockley 'The path to the conception of the junction transistor' (1976) p. 603

32 G. L. Pearson and W. H. Brattain, 'History of semiconductor research', *Proc. IRE,* **43** (1955) 1799

33 Walter Brattain, 'Genesis of the transistor', *The Physics Teacher* (March 1968) 110

34 G. L. Pearson and W. H. Brattain, 'History of semiconductor research', p. 1799

35 Walter Brattain, 'Genesis of the transistor'; interview with Walter Brattain; 'Biography of Dr William Shockley', *Solid State Journal* (May 1961)

36 Interview with Ralph Bray

37 Interview with V. A. Johnson

38 *Ibid.*

39 Interview with Walter Brattain

40 John Bardeen, 'A whole series of breakthroughs', *Electronics and Power* (22 February 1973) 57

41 Jack Morton, 'From research to technology', *International Science and*

Technology, **29** (May 1964) 91; Charles Weiner, 'How the transistor emerged', pp. 25–6

42 Interview with Walter Brattain; W. S. Gorton, 'Memorandum for record. The Genesis of the transistor – case 38139–8' (Niels Bohr Library, American Institute of Physics, New York – tyepscript) p. 1

43 C. A. Hogarth, 'The transistor – its invention and its current prospects', *Physics in Technology,* **4,** 3 (1973) 178

44 Interview with Walter Brattain

45 William Shockley, 'The invention of the transistor' (1972) p. 58

46 *Ibid.,* p. 63; W. S. Gorton, 'Memorandum for record' pp. 1–2

47 W. S. Gorton, 'Memorandum for record', p. 1; William Shockley, *The Invention of the Transistor* (1974) p. 55

48 Richard Nelson, 'The link between science and invention: the case of the transistor', in U.S. Bureau of Economic Research, *The Rate and Direction of Inventive Activity – Economic and Social Factors,* Princeton University Press, Princeton (1962) pp. 561–2

49 William Shockley, 'The invention of the transistor' (1972) p. 63

50 Walter Brattain, 'Genesis of the transistor'; W. H. Brattain, *Experiments Leading Up to the Discovery of the Transistor,* transcript of Bell Conference, Murray Hill (27 October 1948); W. S. Gorton, 'Memorandum for record'

51 Shockley has written at length on what he calls 'creative-failure methodology'. William Shockley, 'The invention of the transistor' (1972) and *The Invention of the Transistor* (1974)

52 See, e.g. W. H. Brattain, 'How the transistor was named (Niels Bohr Library, American Institute of Physics, New York – typescript) p. 1; John Bardeen, 'Background of transistor development', in *Physics 50 Years Later,* National Academy of Science, Washington (1973) p. 176

53 Interviews with Walter Brattain and R. W. Pohl

54 Interview with Walter Brattain

55 Ray Connolly, 'Does Bell Labs live up to reputation?', p. 51

56 Interview with Harvey Brooks

57 Gordon Teal, 'Roots of creative research', *Idea,* **11** (Spring 1965) 3; Michael Wolff, 'The R & D bootleggers: inventing against odds', *IEEE Spectrum* (Spring 1975) 38–42

58 Jack Morton, 'From research to technology', pp. 87–8

59 Interview with Walter Brattain

60 Interview with James M. Early

61 Charles Weiner, 'How the transistor emerged', pp. 28–9

62 Daniel M. Costigan, 'The quest for the crystal that amplifies', *Popular Electronics* (June 1968) p. 43; interview with James M. Early

63 Interview with C. A. Hogarth

64 William Shockley, *Electrons and Holes in Semiconductors,* Van Nostrand, New York (1950)

65 Richard Nelson, 'The link between science and invention', pp. 564–6

66 William Shockley, 'The path to the conception of the junction transistor' (1976) p. 599

Chapter 5

1 C. A. Hogarth, 'The transistor – its invention and its current prospects', *Physics in Technology,* **4,** 3 (1973) 180
2 Interview with D. A. Wright
3 Interview with Dieter Alsberg
4 William Shockley, *The Invention of the Transistor – An Example of Creative-Failure Methodology,* National Bureau of Standards Publication No. 388 (May 1974) p. 74
5 William Shockley, 'The invention of the transistor: an example of creative-failure methodology', *Solid State Devices,* Institute of Physics, London (1972) p. 64
6 Interview with J. R. Pierce
7 Walter Brattain, 'How the transistor was named', (Niels Bohr Library, American Institute of Physics – typescript); Walter Brattain, 'Genesis of the transistor', *The Physics Teacher* (March 1968) 113
8 Interview with Walter Brattain
9 E. Aisberg, 'Transitron = transistor + ?' *Toute la Radio,* **16,** 137 (July/August 1949) 218–20; interview with Frederick Seitz
10 Interview with John Tilton
11 J. Bardeen and W. H. Brattain, 'The transistor, a semiconductor triode', *Physical Review* (15 July 1948) 230–1. The announcement acknowledged the help of William Shockley and others at Bell Laboratories.
12 Interview with Walter Brattain
13 'Fifth anniversary of transistor announcement' *Bell Laboratories Record,* **31** (July 1953) 276–8; U.S. Army Electronics Command, *Miniaturization and Microminiaturization of Army Communications – Electronics, 1946–64,* Historical Monograph *L,* Fort Monmouth, New Jersey (1965) p. 116
14 Interviews with Herbert Kleiman and John Hallowes
15 Interview with Jerome Kraus
16 Interview with Dieter Alsberg
17 'The transistor', *Bell Laboratories Record,* **26,** 8 (August 1948) 321–4
18 Mervin Kelly, 'The first five years of the transistor', *Bell Telephone Magazine,* **32** (1953) 74; Charles Weiner, 'How the transistor emerged', *IEEE Spectrum* (January 1973) 25
19 *New York Times* (1 July 1948)
20 See Stuart MacDonald and Ernest Braun, 'The transistor and attitude to change', *American Journal of Physics,* November 1977
21 C. A. Hogarth, 'The transistor', p. 180
22 E.g. Morgan Sparks, '25 years of transistors', *Bell Laboratories Record* (December 1972) 342
23 Ralph Bown, 'The transistor as an industrial research episode', *Scientific Monthly,* **80** (January 1955) 42
24 E.g. 'Inch-high substitute for vacuum tube', *Science Illustrated,* **3** (October 1948) 69
25 E.g. W. A. Wildhack, 'Transistor', *Review of Scientific Instruments,* **19** (October 1948) 724; 'The transistor – a crystal triode', *Electronics*

(September 1948) 68

26 *Telegraph and Telephone Age,* **66,** 8 (August 1948) 7–30

27 *Product Engineering,* **19,** 9 (September 1948) 158–9

28 'Crystal amplifier', *Science News,* **10** (1949) 135

29 'Transistor may replace vacuum tubes', *Tele-Tech,* **7,** 8 (August 1948) 19

30 'Fifth anniversary of transistor announcement', *Bell Laboratories Record,* **31** (July 1953) 278

31 John E. Tilton, *International Diffusion of Technology. The Case of Semiconductors,* Brookings Institution, Washington D.C. (1971) p. 75

32 Interview with Gunther Rudenberg

33 Morgan Sparks, '25 years of transistors', pp. 343–4

34 Ralph Bown, 'Inventing and patenting at Bell Laboratories', *Bell Laboratories Record,* **32** (January 1954) 9; interview with Gunther Rudenberg

35 'The improbable years', *Electronics,* **41** (19 February 1968) 81; interview with Jerome Kraus

36 M. D. Fagen (ed.), *Impact,* Bell Laboratories, New Jersey (1972) p. 56

37 Francis Bello, 'The year of the transistor', *Fortune* (March 1953) 132

38 *Ibid.,* pp. 129, 132

39 James Fahnestock, 'Transistorized hearing aids', *Electronics,* **26** (April 1953) 154–5

40 'Government hearing aids', *Wireless World,* **54** (January 1948) 11–12; 'Transistors miniaturize hearing aids', *Electrical Manufacturing,* **51,** 3 (March 1953) 146–8; 'Hearing aids switch to transistors', *Tele-Tech,* **12,** 3 (March 1953) 158; interview with Everitt Coon

41 'Transistors miniaturize hearing aids', p. 145

42 S. F. Lybarger, 'A discussion of hearing aid trends', *International Audiology,* **5** (1966) 376–83; William Greenbaum, 'Miniature radio amplifiers', *Journal of the Audio Engineering Society,* **15,** 4 (October 1967) 438–44

43 Joseph O'Connor, 'What can transistors do?', *Chemical Engineering,* **59** (May 1952) 156

44 Interviews with J. R. Tillman, Clare Thornton and J. R. Pierce

45 Bell Laboratories, *The Transistor. Selected Reference Material on Characteristics and Applications,* New York (1951) p. 2

46 Interview with Alan Gibson

47 'Transistors: growing up fast', *Business Week* (5 February 1955) 86

48 W. F. Starr, 'Availability of transistors', *Progress in Quality Electronic Components,* IRE–AIEE–RTMA Symposium, Washington D.C. (5–7 May 1952) pp. 150–1

49 Sir John Cockcroft, 'The process of technological innovation', *Proceedings of IERE,* **3,** 3 (July/August 1965) 91

50 Interviews with George Abraham and J. R. Pierce

51 E.g. C. F. Carter and B. F. Williams, *Industry and Technical Progress,* Oxford University Press, London (1957) pp. 22–3

52 E.g. 'The transistor', *Electrical News and Engineering,* **58,** 18 (September 1949) 89–92, 94; interview with Gunther Rudenberg

53 Walter Richter, 'Industrial electronic developments in the last two decades and a glimpse into the future', *Proceedings of IRE,* **50** (1962) 1138

54 E.g. S. Young White, 'Experimental germanium crystal amplifier', *Audio Engineering* (August 1948,) 28–9, 38–9; C. E. Atkins, 'A crystal that amplifies', *Radio and Television News,* **40** (October 1948) 39, 181–4

55 Donald G. Fink, 'The decline and fall of the free electron', *Proceedings of the National Electronics Conference,* vol. IV (Chicago, 4–6 November 1948) p. 5

56 Joseph O'Connor, 'What can transistors do?'

57 E.g. Mervin Kelly, 'The first five years of the transistor', pp. 85–6; Mervin Kelly, 'The first decade of the transistor', *Bell Telephone Magazine,* **37** (1958) 38

58 *Nobel Lectures, Physics, 1942–62,* Elsevier, Amsterdam (1964) pp. 313–86

59 E. Jantsch, *Technological Planning and Social Futures,* Cassell/Associated Business Programmes, London (1972) p. 115

60 Interview with Quentin Kaiser

61 Interviews with Alan Gibson and Ralph Bray

62 A. F. Gibson, *Progress in Semiconductors,* Heywood, London (1956) p. 4. See also E. E. David Jr, 'The role of research', *Bell Laboratories Record,* **47** (January 1969) 9

63 Alan Gibson in *Report of the International Conference on the Physics of Semiconductors,* Institute of Physics and The Physical Society, London (1962) p. 901

Chapter 6

1 Arthur D. Little, *Patterns and Problems of Technical Innovation in American Industry, Part IV: The Semiconductor Industry,* Report PB 181573 to National Science Foundation, Washington D.C. (1963) p. 152

2 'The trend of affairs', *Technology Review,* **55,** 3 (January 1953) 144

3 C. A. Hogarth, 'The transistor – its invention and its current prospects', *Physics in Technology,* **4,** 3 (1973) 181

4 J. J. Sparkes, 'The first decade of transistor development: a personal view', *Radio and Electronic Engineer,* **43,** 1/2 (January/February 1973)4

5 W. F. Starr, 'Availability of transistors', *Progress in Quality Electronic Components,* IRE–AIEE–RTMA Symposium, Washington D.C. (5–7 May 1952) p. 151

6 Arthur D. Little, *Patterns and Problems of Technical Innovation in American Industry, Part IV,* pp. 152–3

7 B. N. Slade, 'Survey of transistor development – Part 2', *Radio and Television News,* **48,** 10 (October 1952) 115–6

8 Arthur D. Little, *Patterns and Problems of Technical Innovation in American Industry, Part IV,* p. 152

9 J. J. Sparkes, 'The first decade of transistor development', p. 4

10 *Ibid.,* pp. 4–5

11 A. M. Golding, 'The semiconductor industry in Britain and the United

States: A case study in innovation, growth and the diffusion of technology' (D.Phil. thesis, University of Sussex, 1972) pp. 72–3

12 Gordon Teal, 'Roots of creative research', *Idea,* **11** (Spring 1965) 4–5

13 A. M. Golding, 'The semiconductor industry in Britain and the United States', p. 73

14 *Ibid.*, p. 74

15 R. M. Ryder, 'Ten years of transistors', *Radio-Electronics* (May 1958) 35–6; H. S. Blanks, 'Transistor trends and developments', *Proceedings of IRE of Australia,* **19** (July 1958) 379

16 'The trend of affairs', p. 144

17 J. A. Morton, 'The technological impact of transistors', *Proceedings of IRE,* **46** (1958) 959

18 T. R. Scott and S. E. Meyer, 'Engineering and chemical aspects of semiconductors', in H. K. Henisch (ed.) *Semi-Conducting Materials,* Butterworths, London (1951) p. 254; Frederick Seitz, 'Solid', *Physics Today,* **2**, 6 (1949) 22

19 Richard L. Petritz, 'Contributions of materials technology to semiconductor devices', *Proceedings of IRE,* **50** (1962) 1026

20 Michael Wolff, 'The R & D bootleggers: inventing against odds', *IEEE Spectrum* (July 1975) 39–42

21 Interview with Alan Gibson

22 Mervin J. Kelly, 'The first decade of the transistor', *Bell Telephone Magazine,* **37** (Summer 1958) 29

23 Arthur D. Little, *Patterns and Problems of Technical Innovation in American Industry, Part IV*, pp. 153–4; Richard L. Petritz, 'Contributions of materials technology to semiconductor devices,' p. 1027; R. M. Ryder 'Ten years of transistors', *Radio Electronics* (May 1958) 35–6

24 Richard L. Petritz, 'Contributions of materials technology to semiconductor devices', pp. 1025, 1028; Mervin J. Kelly, 'The first decade of the transistor', p. 30; Gordon Teal, 'Roots of creative research', p. 5

25 H. S. Blanks, 'Transistor trends and developments'

26 Mervin J. Kelly, 'The first decade of the transistor', p. 30

27 H. S. Blanks, 'Transistor trends and developments', p. 378

28 C. A. Hogarth, 'The transistor – its invention and its current prospects', p. 180

29 B. N. Slade, 'Survey of transistor development', *Radio and Television News* (September 1952) 44; 'GE predicts germanium will miniaturize electronic components', *Product Engineering,* **23** 3 (1952) 179

30 R. W. Douglas and E. G. James, 'Germanium in electronics', *Times Review of Industry* (November 1948) 21

31 R. M. Ryder 'Ten years of transistors', p. 37; Richard L. Petritz, 'Contributions of materials technology to semiconductor devices'

32 E.g. H. S. Blanks 'Transistor trends and developments', p. 380

33 E. G. Bylander, 'Semiconductor materials for high temperatures', *Electro-Technology,* **72**, 3 (1963) 126–7

34 John Tilton, *International Diffusion of Technology. The Case of Semiconductors,* Brookings Institution, Washington D.C. (1971) p. 52

35 *Ibid.,* pp. 49–55
36 *Ibid.,* pp. 60–2
37 Electronic Industries Association, *Electronic Market Data Book,* Washington D.C. (1974) pp. 79, 87, 89
38 *Ibid.,* p. 79
39 Jerome Kraus, An economic study of the U.S. semiconductor industry (Ph.D. thesis, New School for Social Research, 1973) p. 158
40 Interviews with J. R. Pierce and Henry Levinstein
41 John Tilton, *International Diffusion of Technology,* pp. 55, 65
42 'Transistors enter commercial field', *Product Engineering* (March 1953) 197
43 'Transistors: growing up fast', *Business Week* (5 February 1955) 86
44 William B. Harris, 'The battle of the components', *Fortune,* 55 (May 1957) 138
45 John Tilton, *International Diffusion of Technology,* p. 67
46 *Ibid.,* p. 51
47 A. M. Golding, 'The semiconductor industry in Britain and the United States', p. 157
48 P. E. Haggerty, 'Objectives, strategies and tactics' in Texas Instruments, *Management Philosophies and Practices of Texas Instruments Inc.,* Dallas (1965) p. 51
49 Patrick E. Haggerty, 'Strategies, tactics and research', *Research Management,* 9, 3 (1966) 149–51
50 'Research packed with Ph.D.'s', *Business Week* (22 December 1956) 57–8
51 A. M. Golding, 'The semiconductor industry in Britain and the United States', p. 158
52 William B. Harris, 'The company that started with a gold whisker', *Fortune,* 60 (August 1959) 98–101, 136, 140, 142, 145; 'Transitron Electronic Corporation', *Solid State Journal,* 2 (March 1961) 52–5
53 Interview with Alan Goss
54 William B. Harris, 'The battle of the components', p. 286
55 Interview with Alan Gibson
56 Interview with Frederick Seitz
57 Shirley Thomas, *Men of Space,* Chilton, Philadelphia (1962) p. 199
58 William B. Harris, 'The company that started with a gold whisker', p. 136
59 Interview with Alex Kasatkin
60 Interview with Douglas Warschauer
61 Interview with Robert Noyce
62 Interview with Alex Kasatkin
63 Interview with John Rogers
64 *Ibid.*
65 Interview with Everitt Coon
66 Interview with Jack Kilby
67 Interview with Tom Mitchell
68 Interview with Clare Thornton
69 Arthur D. Little, *Patterns and Problems of Technical Innovation in American*

Industry, Part IV, p. 145

70 Interview with J. R. Pierce
71 'Semiconductors', *Business Week* (26 March 1960) 101
72 'Transistors enter commercial field', p. 197
73 'Semiconductors will become a major field of electrical engineering', *Product Engineering,* **22**, 5 (May 1951) 132–3; Stephen J. Angello, 'Semiconductors – new vigor in an old field', *Westinghouse Engineer,* **10** (July 1950) 186
74 Fifth anniversary of transistor announcement', *Bell Laboratories Record,* **31** (July 1953) 276–8
75 'Transistors enter telephone service', *Bell Laboratories Record,* **30** (November 1952) 439
76 Joseph O'Connor, 'What can transistors do?', *Chemical Engineering,* **59** (May 1952) 155
77 Interviews with James M. Early and Billy M. Horton
78 Mervin J. Kelly, 'The first five years of the transistor', *Bell Telephone Magazine,* **32** (Summer 1953) 86; Mervin J. Kelly, 'The first decade of the transistor', pp. 27–8; David D. Holmes, 'Application of transistors in communications equipment', *Proceedings of IRE,* **46** (June 1958) 1255
79 Interview with J. R. Tillman
80 'John Q meets the transistor', *Electronics,* **26**, 1 (January 1953) 6; Francis Bello, 'The year of the transistor', *Fortune* (March 1953) 134
81 Interview with Everitt Coon
82 Francis Bello, 'The year of the transistor', p. 164
83 Joseph O'Connor, 'What can transistors do?', pp. 154, 370
84 'Automatic factory near for electronics', *Aviation Week,* **57**, 17 (1952) 36
85 'Transistors: growing up fast', *Business Week* (5 February 1955) 86
86 H.S. Blanks, 'Transistor trends and developments', p. 377
87 J. N. Barry and A. E. Jackets, 'The application of transistor to television', *Journal of the Television Society,* **8**, 8 (1957) 333–4
88 Louis N. Ridenour, 'A revolution in electronics', *Scientific American,* **185**, 2 (1951) 17; Francis Bello, 'The year of the transistor', p. 162
89 'Transistors: growing up fast', p. 86
90 Interviews with Norman Doctor and Robert Noyce
91 Jerome Kraus, *An Economic Study of the U.S. Semiconductor Industry,* p. 90
92 William B. Harris, 'The battle of the components', p. 137
93 I. R. Obenchain and W. J. Galloway, 'Transistors and the Military', *Proceedings of IRE,* **40** (November 1952) 1287
94 Francis Bello, 'The year of the transistor', p. 132
95 Edwin A. Speakman, 'Reliability of military electronics', *Progress in Quality Electronics Components,* IRE–AIEE–RTMA Symposium, Washington D.C. (5–7 May 1952) pp. 8–9
96 *Ibid.*
97 O. M. Stuetzer, 'Transistor in airborne equipment', *Proceedings of IRE,* **40** (November 1952) 1529–30
98 Richard E. Kimes, 'Transistor decade counter', in *The Application of Transistors to Military Electronics Equipment,* Committee on Electronics

Symposium, Office of Secretary of Defense, Yale (2–3 September 1953) pp. 164–5; see also N. W. Feldman and W. W. Hall, 'Transistor phono-amplifier', p. 194

99 U.S. Army Electronics Command, *Miniaturization and Microminiaturization of Army Communications – Electronics, 1946–1964,* Historical Monograph L, Fort Monmouth, New Jersey, 1965, pp. 116–119

100 John Tilton, *International Diffusion of Technology,* p. 93

101 *Ibid.,* p. 92

102 'Transistors: growing up fast', p. 87

103 Electronic Industries Association, *Electronic Market Data Book,* Washington D.C. (1974) p. 87

104 H. Rice, 'From transistors to microelectronics', Speech to AFCEA Panel, Fort Monmouth, N.J. (1968) p. 3

105 *Ibid.,* p. 4

106 John Tilton, *International Diffusion of Technology,* p. 94

107 Jerome Kraus, 'An Economic Study of the U.S. semiconductor industry', p. 90

108 John Tilton, *International Diffusion of Technology,* p. 91

Chapter 7

1 'Semiconductors', *Business Week* (26 March 1960) 96–101

2 J. J. Sparkes, 'The first decade of transistor development: a personal view', *Radio and Electronic Engineer,* 43, 1/2 (1973) 7–8

3 'Semiconductors', p. 101

4 J. J. Sparkes, 'The first decade of transistor development', p. 8

5 'Semiconductors', p.101

6 Jerome Kraus, 'An economic study of the U.S. semiconductor industry' (Ph.D. thesis, New School for Social Research, 1973) p. 32

7 'The improbable years', *Electronics,* 41 (19 February 1968) 83–4

8 A. M. Golding, *The semiconductor industry in Britain and the United States: a case study in innovation, growth and the diffusion of technology'* (D.Phil. thesis, University of Sussex, 1972) p. 76

9 Personal correspondence with Harry Diamond Laboratories

10 M. Smollet, 'The technology of semiconductor manufacture', *Radio and Electronic Engineer,* 43, 1/2 (1973) 30–2

11 J. Kosar, *Light Sensitive Systems,* John Wiley, London (1965)

12 A. M. Golding, 'The semiconductor industry in Britain and the United States', pp. 160–1

13 J. J. Sparkes, 'The first decade of transistor development', pp. 8–9; K. J. Dean and G. White, 'The semiconductor story, 2', *Wireless World,* 79, 1448 (1973) 67–8; A. M. Golding, 'The semiconductor industry in Britain and the United States', p. 77

14 Robert E. Freund, 'Competition and innovation in the transistor industry' (Ph.D. thesis, Duke University, 1971) pp. 69–70

15 Jerome Kraus, 'An economic study of the U.S. semiconductor industry', pp. 206–211

16 Jerry Eimbinder, 'Transistor industry growth patterns', *Solid State Design,* **4** (January 1963) 57–9

17 Jerome Kraus, 'An economic study of the U.S. semiconductor industry', p. 145

18 William Shockley, 'An invited essay on transistor business', *Proceedings of the IRE.,* **46**, 6 (1958) 954–5

19 Jerome Kraus, 'An economic study of the U.S. semiconductor industry', p. 95

20 *Ibid.,* p. 91

21 'Our growing components market', *Electronics* (5 January 1962) 66–7

22 'Semiconductors', p. 106

23 William B. Harris, 'The company that started with a gold whisker', *Fortune,* **60** (August 1959) 100

24 'Transitron sets investors agog', *Business Week* (5 December 1959) 123–4

25 Jerome Kraus, 'An economic study of the U.S. semiconductor industry', p. 98

26 Barry Miller, 'Semiconductor crisis looms over pricing', *Aviation Week* (12 June 1961) 73–85; Barry Miller, 'Competition to tighten in semiconductors', *Aviation Week & Space Technology* (23 September 1963) 56–68; Arthur D. Little, *Patterns and Problems of Technical Innovation in American Industry, Part IV: The Semiconductor Industry,* Report PB 181573 to National Science Foundation, Washington D.C. (1963) pp. 163–4; A. M. Golding, 'The semiconductor industry in Britain and the United States', pp. 132–3

27 Patrick Conley, 'Experience curves as a planning tool', *IEEE Spectrum* (June 1970) 22–8

28 'Texas Instruments: "all systems go" ', *Dun's Review,* **89** (January 1967) 29

29 E.g. John Tilton, *International Diffusion of Technology. The Case of Semiconductors,* Brookings Institution, Washington D.C. (1971) 85–87

30 'Semiconductors', p. 101

31 Robert E. Freund, 'Competition and innovation in the transistor industry', p. 52

32 Herbert Kleiman, 'The integrated circuit: a case study of product innovation' (D.B.A. thesis, George Washington University, 1966) pp. 158–60

33 *Ibid.,* pp. 161–5

34 Barry Miller, 'Semiconductor crisis looms over pricing', p. 75

35 Lee Weddig, '20 per cent of transistor market eyed via Motorala push', *Electronic News* (6 May 1957) 1, 19

36 Robert Henkel, 'Motorola Div. sales seen close to tripling '59 level', *Electronic News* (5 September 1960) 22

37 Electronics Industries Association, *Electronic Market Data Book,* Washington D.C. (1974)

38 Edward Ney Dodson III, 'Component product flows in the electronics industry' (Ph.D. thesis, Stanford University, 1966)

39 William Long II, 'Price and nonprice practices under the uncertain

conditions of rapidly improving technologies – a case study' (Ph.D. thesis, George Washington University, 1967)

Chapter 8

1 A. M. Golding, 'The semiconductor industry in Great Britain and the United States: a case study in innovation, growth and the diffusion of technology' (D.Phil. thesis, University of Sussex, 1972) p. 78. Two recent excellent accounts of the development of integrated circuitry exist in Jack S. Kelly, 'Invention of the integrated circuit', *IEEE Transactions on Electron Devices,* **ED-23,** 7 (July 1976) 648–54; and Michael F. Wolff, 'The genesis of the integrated circuit', *IEEE Spectrum* (August 1976) 45–53

2 Jack A. Morton, 'The microelectronics dilemma', *International Science and Technology,* **55** (July 1966) 37; Jack A. Morton, *Organizing for Innovation,* McGraw-Hill, New York (1971) pp. 106–7

3 'Jack Kilby: The champion of IC's, *Electronic Engineer's Design Magazine* (January 1967)

4 A. M. Golding, 'The semiconductor industry in Great Britain and the United States', p. 79

5 Interview with Jack Kilby

6 A. M. Golding, 'The semiconductor industry in Great Britain and the United States', p. 79

7 'Fairchild scores a point on circuits', *Business Week* (November 1969) 128

8 Interview with Robert Noyce

9 Laurence D. Shergalis, 'Microelectronics – a new concept in packaging'. *Electronic Design* (29 April 1959) 32

10 'Transistors and component miniaturization', *British Communications and Electronics,* **6** (June 1959) 462; G. W. A. Dummer, 'Miniaturization and micro-miniaturization', *Wireless World,* **65** (1959) 545–9

11 L. J. Ward, 'Microminiaturization' *Journal of the Institute of Electronic Engineers,* **8** (1962) 200

12 E.g. 'Research packed with Ph.D.'s', *Business Week* (22 December 1956) 56–64

13 Interview with Quentin Kaiser

14 'Millions of components per cubic foot', *Automatic Control* (September 1956) 6–14. See also G. W. A. Dummer, 'Miniaturization and micro-miniaturization', p. 546; James W. Granville, 'Microminiaturization in electronics', *New Scientist,* **7** (28 April 1960) 1078; Laurence D. Shergalis, 'Microelectronics', pp. 32–49

15 Interview with G. W. A. Dummer

16 William H. Greenbaum, 'Miniature audio amplifiers', *Journal of the Audio Engineering Society,* **15,** 4 (1967) 442

17 Herbert Kleiman, 'The integrated circuit: a case study of product innovation' (D.B.A. thesis, George Washington University, 1966) p. 59

18 'Jack Kilby: The champion of IC's'
19 'Microelectronics today', *Electronic Industries* (December 1962) 92
20 *Ibid.,* p. 93
21 'Army boosts micromodule outlay for '63', *Electronic News* (3 September 1962) 1, 39
22 Herbert Kleiman, 'The integrated circuit', pp. 51–5
23 James W. Granville, 'Microminiaturization in electronics', p. 1077, G. W. A. Dummer, 'Progress with extremely small electronic circuits', *New Scientist,* 325 (7 February 1963) 283–4
24 Herbert Kleiman, 'The integrated circuit', pp. 50–1; Laurence D. Shergalis, 'Microelectronics', p. 32.
25 Interview of personnel of Harry Diamond Laboratories (formerly D.O.F.L.); HDL Disposition Form AMXDO-RDB (20 April 1970)
26 G. W. A. Dummer, 'Electronic components in Great Britain', *Progress in Quality Electronics Components,* Proceedings of the Symposium of the IRE–AIEE–RTMA, Washington D.C. (5–7 May 1952) p. 19
27 G. W. A. Dummer, 'A review of British work on microminiaturization techniques', in *Electronics Reliability and Microminiaturization,* Pergamon, London, vol. 1 (1962,) pp. 39–41
28 Interview with G. W. A. Dummer
29 Interviews with Jack Kilby and Robert Noyce
30 K. J. Dean and G. White, 'The semiconductor story – 3', *Wireless World,* 79, 1499 (1973) 137
31 Interview with G. W. A. Dummer
32 Interview with Gene Strull
33 Herbert Kleiman, 'The integrated circuit', 180–4
34 *Ibid.,* p. 181; interview with Gene Strull
35 'The new shape of electronics', *Business Week* (14 April 1962) 174; interview with William Corak
36 James M. Bridges, 'Integrated electronics in defence systems', *Proceedings of the IEEE,* 52, 12 (1964) 1407
37 Interview with Gene Strull
38 Herbert Kleiman, 'The integrated circuit', p. 185
39 'Jack Kilby: The champion of IC's'
40 'Army boosts micromodule outlay for '63'
41 Interview with Norman Doctor
42 Herbert Kleiman, 'The integrated circuit', p. 108
43 E.g. '1964: The year microcircuits grew up', *Electronics,* 37, 11 (1964) 10–11
44 'Millions of components per cubic foot', *Automatic Control* (September 1958) 6; 'Our growing components market', *Electronics* (5 January 1962) 67–8
45 L. J. Ward, 'Microminiaturisation', *Journal of IEE,* 8 (1962) 201
46 G. W. A. Dummer, 'A review of British work on microminiaturization techniques', p. 41
47 James Kendall, 'Electronic engineering in 1973', *World Aviation Electronics* (January 1963) 15–19; John R. Riggs, 'New development to

achieve old goals', *Electrical Manufacturing*, **65**, 5 (May 1960) 263

48 Interview with Jack Kilby

49 'Semiconductor crisis looms over pricing', *Aviation Week* (12 June 1961) 84–5

50 Jack A. Morton, 'The microelectronics dilemma', pp. 36–7; 'The new shape of electronics', p. 168; J. A. Morton, 'From physics to function', *IEEE Spectrum*, **2** (September 1965) 62

51 'The new shape of electronics', p. 168

52 *Ibid.*

53 Herbert Kleiman, 'The integrated circuit', pp. 118–19

54 James Kendall, 'Electronic engineering in 1973', p. 15

55 Interview with Gunther Rudenberg

56 Barry Miller, 'Competition to tighten in semiconductors', *Aviation Week* (23 September 1963) 63, 65

57 Arthur P. Stern, 'Preface to the integrated electronics issue', *Proceedings of IEEE*, **52**, 12 (1964) 1395

58 'Boom or bomb?', *Electronics* (7 September 1964) 15; J. G. Linvill, J. B. Angell and R. L. Pritchard, 'Integrated electronics vs electrical engineering education', *Proceedings of IEEE*, **52**, 12 (1964) 1425–9

59 Interview with Herbert Kleiman

60 P. E. Haggerty, 'Integrated electronics – a perspective', *Proceedings of IEEE*, **52**, 12 (1964) 1404

61 Interviews with Richard Gerdes and Alan Esbitt

62 E.g. Patrick E. Haggerty, 'The economic impact of integrated circuitry', *IEEE Spectrum*, **1** (June 1964) 80–2

63 Barry Miller, 'Report sees semiconductor dominance', *Aviation Week and Space Technology* (10 December 1962) 95–6

64 Electronic Industries Association, *Market Data Book,* Washington D.C. (1974) p. 1; John Tilton, *International Diffusion of Technology. The Case of Semiconductors,* Brookings Institution, Washington, D.C. (1971) p. 91

65 Barry Miller, 'Study forecasts microcircuitry growth', *Aviation Week and Space Technology* (25 February 1963) 84–96

66 A. M. Golding, 'The semiconductor industry in Great Britain and the United States', p. 83

67 Herbert Kleiman, 'The integrated circuit', pp. 128–9

68 'MOS – the price race', *Electronic News* (29 September 1969) 1, 51; 'An integrated circuit that is catching up', *Business Week* (25 April 1970) 134–6

69 David Hester, 'Name of the game is LSI', *New Scientist* (14 November 1974) 508–9

70 A. M. Golding, 'The semiconductor industry in Great Britain and the United States', p. 55

71 'Where the action is in electronics', *Business Week* (4 October 1969) 97

72 Patrick E. Haggerty, 'The role of institutional culture in an industrial corporation', Leatherbee Lecture to Harvard Graduate School of Business Administration (11 March 1975) p. 8

73 A. M. Golding, 'The semiconductor industry in Great Britain and the

United States', pp. 53–4
74 'Where the action is in electronics' p. 97
75 Jerome Kraus, *An Economic Study of the U.S. Semiconductor Industry* (Ph.D. thesis, New School for Social Research, 1973) p. 56
76 Interviews with Kurt Hoselitz and Robert Stratton
77 K. J. Dean and G. White, 'The semiconductor story – 4', *Wireless World,* **79**, 1450 (1973) 171; J. A. Morton, *Organizing for Innovation,* pp. 110–13
78 'IC makers' surge to continue', *Electronics,* **24**, 1 (1969) 117
79 David Hester, 'Name of the game is LSI', p. 509
80 'Where time moves at a dizzying pace', *Business Week* (20 April 1968) 182
81 'MOS – the price race', p. 51
82 William Finan, *International Transfer of Semiconductor Technology Through U.S.-Based Firms,* National Bureau of Economic Research, New York (1975)
83 G. W. A. Dummer, 'Integrated electronics – a historical introduction', *Electronics and Power* (March 1967); S. H. Hollingdale and G. C. Toothill, *Electronic Computers,* Penguin, London (1975) p. 350
84 Electronic Industries Association, *Market Data Book,* Washington D.C. (1975)

Chapter 9

1 Jerome Kraus, 'An economic study of the U.S. semiconductor industry' (Ph.D. thesis, New School for Social Research 1973) pp. 118, 120
2 Robert Freund, 'Competition and innovation in the transistor industry' (Ph.D. thesis, Duke University, 1971) p. 25
3 John Tilton, *International Diffusion of Technology. The Case of Semiconductors,* Brookings Institution, Washington D.C. (1971) p. 57
4 William Finan, *The International Transfer of Semiconductor Technology Through U.S.-Based Firms,* National Bureau of Economic Research, New York (1975) p. 11
5 *Ibid.,* p. 5
6 *Ibid.,* p. 7
7 Jerome Kraus, 'An economic study of the U.S. semiconductor industry', p. 129
8 R. S. Estall, 'The electronics products industry of New England', *Economic Geography,* **39** (July 1963) 189–216
9 Elizabeth Deutermann, 'Seeding science-based industry', *Federal Reserve Bank of Philadelphia Business Review* (May 1966) pp. 3–10
10 Interview with Tom Mitchell
11 Nilo Lindgren, 'The splintering of the solid-state electronics industry', in *Dealing with Technological Change,* Auerbach/Innovation Magazine, New York (1971) p. 34
12 'Semiconductor family tree', *Electronic News* (8 July 1968) 4–5, 38
13 'Biography. Dr. William Shockley', *Solid State Journal* (May 1961)
14 Shirley Thomas, *Men of Space,* Chilton Publishing, Philadelphia (1962) pp. 199–200

15 Nilo Lindgren, 'The splintering of the solid-state electronics industry', p. 39
16 Interview with Robert Noyce
17 Nilo Lindgren, 'The splintering of the solid-state electronics industry', p. 50
18 'Silicon Summit', *Electronic News* (29 September 1969) 1
19 Interview with Phil Ferguson; 'Semiconductor family tree', *Electronic News* (8 July 1968) 4; A. M. Golding, 'The semiconductor industry in Britain and the United States: a case study in innovation, growth and the diffusion of technology' (D.Phil. thesis, University of Sussex, 1972) p. 249
20 Interview with Phil Ferguson
21 'Latecomer', *Forbes* (1 June 1969) 24
22 Interview with Tom Mitchell
23 Interview with Bob Cook
24 Nilo Lindgren, 'The splintering of the solid-state electronics industry', p. 49
25 Correspondence with Richard Blanchard, Foothill College, Los Altos, California
26 Interview with Robert Noyce
27 Nilo Lindgren, 'The splintering of the solid-state electronics industry', p. 49
28 Interview with John Rogers
29 Nilo Lindgren, 'The splintering of the solid-state electronics industry', p. 49
30 Interview with James M. Early
31 Interview with Floyd Kvamme
32 Interview with Frank Herman
33 Interview with James M. Early
34 Interviews with Phil Ferguson and Richard Gerdes
35 Interviews with Gunther Rudenberg and Robert Stratton
36 Nilo Lindgren, 'The splintering of the solid-state electronics industry', pp. 50–1
37 Interview with Mike Callahan
38 Interview with Bob Cook
39 Interview with Clare Thornton
40 Clarence H. Danhof, 'Technology transfer by people transfer: a case study', Discussion Paper 403, Program of Policy Studies in Science and Technology, George Washington University (August 1969)
41 A. M. Golding, 'The semiconductor industry in Britain and the United States', p. 297
42 Interviews with Herbert Kleiman and Bob Cook
43 Interview with Tom Mitchell
44 'The fight that Fairchild won', *Business Week* (5 October 1968) 113; 'Levy ends red alert at Motorola', *Electronic News* (14 October 1968) 51
45 'High noon' *Forbes* (15 February 1971) 26
46 'Logan set to market monolithic memories', *Electronic News* (29

September 1969) 43

47 'Intersil: upstart with talent', *Business Week* (12 September 1970) 74, 76

48 Interview with William Winter

49 'Semiconductors', *Business Week* (26 March 1960) 110

50 Interview with Robert Stratton

51 Interview with Tom Mitchell

52 Interview with Alex Kasatkin

53 'Semiconductor family tree', p. 4

54 'The fight that Fairchild won', p. 106

55 'Where the action is in electronics', *Business Week* (4 October 1969) 90

56 A. M. Golding, 'The semiconductor industry in Britain and the United States', p. 260; Robert Freund, 'Competition and innovation in the transistor industry', p. 73

57 'Transitron sets investors agog', *Business Week* (5 December 1959) 124; 'Semiconductor family tree', p. 4

58 Robert Freund, 'Competition and innovation in the semiconductor industry', p. 26

59 'Semiconductor family tree', p. 4

60 Interviews with Alex Kasatkin and Herbert Kleiman

61 Interview with Gunther Rudenberg

62 A. M. Golding, 'The semiconductor industry in Britain and the United States', p. 255

63 Barry Miller, 'Competition to tighten in semiconductors', *Aviation Week and Space Technology* (23 September 1963) 59; A. M. Golding, 'The semiconductor industry in Britain and the United States', pp. 175, 248

64 'The fight that Fairchild won', p. 114

65 Robert Freund, 'Competition and innovation in the semiconductor industry', p. 83

66 'Philadelphia story', *Electronic News* (15 February 1971) 1, 64; Jerome Kraus, 'An economic study of the U.S. semiconductor industry', pp. 53–63; interview with Gene Strull

67 Interview with Rudolph Verderber

68 Jerome Kraus, 'An economic study of the U.S. semiconductor industry', p. 118

69 A. M. Golding, 'The semiconductor industry in Britain and the United States', p. 83

70 *Ibid.*, p. 251

71 Glen Madland, 'No. 2 – and trying harder', *Electronic Products* (February 1968) 12

72 A. F. Gibson, *Progress in Semiconductors,* Heywood, London (1956)

73 A. F. Gibson in *Report of the International Conference on the Physics of Semiconductors,* Institute of Physics and the Physical Society, London (1962) p. 901

74 John Tilton, *International Diffusion of Technology,* p. 61; A. M. Golding, 'The semiconductor industry in Britain and the United States', p. 134

75 Correspondence with William Finan

76 Interview with Bob Cook

77 Interview with William Finan
78 William B. Harris, 'The company that started with a gold whisker', *Fortune* (August 1959) 142
79 Interview with Alan Esbitt
80 Interviews with Clare Thornton and Richard Gerdes
81 Interviews with Clare Thornton, Suhael Ahmed and Jack Kohn
82 Interview with John Rogers
83 'Semiconductors', *Business Week* (26 March 1960) 103
84 Joe McLean, 'Philco dropping transistors', *Electronic News* (9 September 1963) 1; Joe McLean, 'Philco action on transistors evaluated', *Electronic News* (9 September 1963) 1, 33; A. M. Golding, 'The semiconductor industry in Britain and the United States', pp. 167–8
85 Barry Miller, 'Semiconductor crisis looms over pricing', *Aviation Week* (12 January 1961) 77; Barry Miller, 'Competition to tighten in semiconductors', p. 59
86 Interview with William Finan
87 Interview with Tom Mitchell
88 Interview with Herbert Kleiman
89 'The semiconductor industry: madness? or method?', *Forbes* (15 February 1971) 20–6
90 Max Shapiro, 'The great crash in growth stocks', *Dun's Review* (January 1971) 30–2
91 Interview with William Winter; Jerome Kraus, 'An economic study of the U.S. semiconductor industry', pp. 98–9; interview with William Corak.

Chapter 10

1 Robert Freund, 'Competition and innovation in the transistor industry' (Ph.D. thesis, Duke University, 1971) pp. 79–80
2 John Tilton, *International Diffusion of Technology. The Case of Semiconductors,* Brookings Institution, Washington D.C. (1971) p. 34
3 William Finan, *The International Transfer of Semiconductor Technology Through U.S.-Based Firms,* National Bureau of Economic Research, New York (1975) p. 116
4 For an early view of this problem see Tim Johnson, 'Transistors – still a backward industry', *Statist* (6 September 1963) 694–5
5 William Finan, *The International Transfer of Semiconductor Technology,* pp. 93, 115
6 Interview with Stuart Pettingill
7 O.E.C.D., *Gaps in Technology – Electronic Components,* Paris (1968) pp. 114–15
8 Interview with William Finan
9 William Finan, *The International Transfer of Semiconductor Technology,* pp. 93–6
10 *Ibid.,* pp. 46–7

11 *Ibid.*, pp. 43, 49
12 Interview with Andrew Hassell
13 William Finan, *The International Transfer of Semiconductor Technology*, pp. 50–2
14 William Finan, *The International Transfer of Semiconductor Technology*, p. 119. See also Michael Payne, ' "The American Challenge" on a chip', *Electronics*, 42, 2 (1969) 74–8
15 William Finan, *The International Transfer of Semiconductor Technology*, p. 64
16 Y. S. Chang, *The Transfer of Technology: Economics of Offshore Assembly. The Case of the Semiconductor Industry*, College of Business Administration, Boston University, UNITAR (1971) pp. 26–7
17 *Ibid.*, p. 17; William Finan, *The International Transfer of Semiconductor Technology*, pp. 62–3
18 'Among the Navajo', *Electronics*, 42, 6 (1969) 53
19 William Finan, *The International Transfer of Semiconductor Technology*, p. 63
20 *Ibid.*, p. 87
21 E.g. Y. S. Chang, *The Transfer of Technology*, pp. 29–30
22 Jerome Kraus, 'An Economic study of the U.S. semiconductor industry' (Ph.D. thesis, New School for Social Research, 1973) pp. 238–43
23 Y. S. Chang, *The Transfer of Technology*, pp. 31, 35
24 *Ibid.*, pp. 54–5
25 John Tilton, *International Diffusion of Technology*, pp. 115, 144
26 *Ibid.*, pp. 98–117. The creation of a new Silicon Valley in Scotland was somewhat optimistically declared in Iain Macdonald, 'No slackening in electronics boom', *Financial Times* (18 November 1968) p. 19. See also A. M. Golding, 'The semiconductor industry in Britain and the United States: a case study in innovation, growth and the diffusion of technology' (D.Phil. thesis, University of Sussex, 1972) pp. 230–7
27 'More semiconductor manufacture in Britain', *British Communications and Electronics*, 4 (December 1957) 751
28 O.E.C.D., *Gaps in Technology*, p. 82
29 Interview with Alex Kasatkin
30 O.E.C.D., *Gaps in Technology*, pp. 88–90
31 Interview with D. A. Wright
32 Interview with Alan Gibson and correspondence with J. B. Gunn
33 Interview with G. W. A. Dummer
34 Interview with William Corak
35 Interviews with D. A. Wright and Alan Gibson
36 O.E.C.D., *Gaps in Technology*, pp. 69–70; C. C. Gee, 'World trends in semiconductor developments and production', *British Communications and Electronics*, 6 (June 1959) 450–61
37 John Tilton, *International Diffusion of Technology'* p. 129; O.E.C.D., *Gaps in Technology*, pp. 57–8
38 See R. G. Atterbury, 'Microelectronics – the key to British prosperity', *Electronic Components*, 7 (May 1966) 467–70 and I. Maddock, 'The task of the circuit designer', *Financial Times* (14 May 1968) p. iv

39 John Tilton, *International Diffusion of Technology*, pp. 156–7
40 *Ibid.*, p. 143
41 Interviews with John Tilton and Phil Ferguson
42 John Tilton, *International Diffusion of Technology*, pp. 146–7
43 J. A. Powell, 'The need for Britain to rationalise', *Financial Times* (12 December 1966) p. x, Michael Gunton, 'Britain's battle for electronic independence', *Electronic News* (30 September 1968) 4–5; interviews with Alan Goss and Robert Noyce
44 Christopher Layton, *Ten Innovations*, George Allen & Unwin, London (1972) pp. 104–5, 115–17; J. C. Akerman, 'Industrial opportunities in semiconductors', *Financial Times* (12 December 1966) p. vi
45 Interview with Phil Ferguson
46 Interview with Richard Gerdes
47 O.E.C.D., *Gaps in Technology*, pp. 75–6; interview with G. W. A. Dummer
48 Interview with Robert Noyce
49 O.E.C.D., *Gaps in Technology'* pp. 102–8; A. M. Golding, 'The semiconductor industry in Britain and the United States', pp. 231–2
50 Christopher Layton, *Ten Innovations*, p. 112; Tim Johnson, 'Transistors – still a backward industry', pp. 694–5
51 Interviews with G. W. A. Dummer and Mike Callahan

Chapter 11

1 Interview with Floyd Kvamme
2 Nicholas Valéry, 'The electronic slide rule comes of age', *New Scientist*, 65, 938 (1975) 506–11; Robin Bradbeer, 'The best price is not always the cheapest', *New Scientist* (Calculator Supplement), 68, 975 (1975) pp. x–xiv
3 Nicholas Valéry, 'Coming of age in the calculator business', *New Scientist* (Calculator Supplement), 68, 975 (1975) pp. ii–iv; 'Electronic calculator', *Wireless World*, 78, 1442 (1972) 357.
4 Interview with Suhael Ahmed
5 'The semiconductor industry: madness? or method?', *Forbes* (15 February 1971) 22; 'An integrated circuit that is catching up', *Business Week* (25 April 1970) 136
6 Interview with Jerome Kraus
7 Interview with Suhael Ahmed; Nicholas Valéry, 'Coming of age in the calculator business', p. ii, 'The semiconductor industry: madness? or method?', p. 25
8 Interviews with Robert Noyce, Frank Herman and Herbert Kleiman
9 Interview with Suhael Ahmed
10 *Ibid.*
11 *Ibid.*
12 'Electronic mathematics', *The Guardian* (25 March 1975) p. 10
13 Interview with Stuart Pettingill
14 Nicholas Valéry, 'The electronic slide-rule comes of age', pp. 506–11

15 Interview with Harold Levine
16 Nicholas Valéry, 'What the users think', *New Scientist* (Calculator Supplement), **68**, 975 (1975) p. xv
17 John Lewis, 'The calculator threat to numeracy', *New Scientist* (Calculator Supplement), **68**, 975 (1975) p. xv
18 *Ibid.*, pp. xv–xvi; 'The calculating new world of tiny tots', *Daily Express* (7 February 1975) p. 8
19 Interviews with John Rogers, Gunther Rudenberg, Jerome Kraus and Richard Gerdes
20 Coleman & Company estimates, 1975
21 *Which?* (May 1976) p. 115
22 Interview with Suhael Ahmed
23 Interviews with Alex Kasatkin, Herbert Kleiman, William Corak and Dean Toombs; Stephen A. Thompson, 'Is CMOS the end of the line?', *Electronic Engineer,* **32**, 1 (1973) 40–1
24 Interview with Robert Noyce
25 Interview with Mike Waltz
26 K. G. Marwing and L. J. Murray, 'Electronic switching systems', in Arthur Garratt, *Penguin Technology Survey,* Penguin, London (1966) pp. 54–7
27 Prescott C. Mabon, *Mission Communications,* Bell Laboratories, New Jersey (1975) p. 83
28 'Electronic telephone exchanges for U.K.', *Wireless World,* **79,** 1449 (March 1973) 112; 'Electronics and telephone exchanges', *Wireless World,* **78**, 1440 (June 1972) 255. See also Hazel Duffy, 'Wrong numbers for Post Office?', *The Guardian* (1 March 1977) p. 16
29 James Martin, *Future Developments in Telecommunications,* Prentice-Hall, New Jersey (1971) pp. 362–4
30 Interview with G. W. A. Dummer
31 Interview with J. E. Flood
32 Interview with John Martin
33 Interview with J. E. Flood
34 Interview with Dieter Alsberg
35 Interview with John Martin
36 Interview with Gunther Rudenberg
37 Interview with John Cave
38 Interview with John Martin
39 Interviews with Gunther Rudenberg, Dean Toombs, Suhael Ahmed and Herbert Kleiman
40 Interview with John Martin
41 Interview with Dieter Alsberg
42 'TV–phone link can fill your home with facts', *The Observer* (30 May 1976) p. 9; Alan Burkitt, 'Teletext arrives on the screen', *New Scientist,* **70**, 1002 (1976) 459–61
43 P. E. Love, *A techno-economic forecast of the microelectronics industry,* unpublished report, Programmes Analysis Unit, Didcot, Berks.; interview with John Cave

44 I. M. Mackintosh, 'Dominant trends affecting the future structure of the semiconductor industry', *Radio and Electronic Engineer,* **43**, 1/2 (1973) 153; *The Guardian* (25 August 1976) p. 18. See also I. M. Mackintosh, 'Semiconductor devices – portrait of a technological explosion', *Radio and Electronic Engineer,* **45**, 10, pp. 515–24

45 'Where the action is in electronics', *Business Week* (4 October 1969) 89

46 Interviews with Herbert Kleiman and Mike Waltz

47 Interviews with Clare Thornton and Dean Toombs

48 Interview with William Winter

49 Interviews with Dean Toombs and Suhael Ahmed

50 Interview with Suhael Ahmed

51 Interviews with Jeffrey T. Hamilton and Suhael Ahmed; 'Microprocessors for cars', *New Scientist,* **70**, 1001 (1976) 419

52 W. E. J. Farvis, 'The semiconductor revolution', *Electronics and Power,* **19** (22 February 1973) 55

53 Ken Garrett, 'Brightening up the dashboard', *New Scientist,* **70**, 1002 (1976) 454–6

54 Interviews with Robert Stratton and Philip Kane; 'Cars begin to bristle with electronics', *New Scientist,* **71**, 1009 (1976) 129

55 S. H. Hollingdale and G. C. Toothill, *Electronic Computers,* Penguin, London (1975) pp. 15–62; Henry Jacobowitz, *Electronic Computers Made Simple,* W. H. Allen, London (1967) pp. 5–10

56 Roger Hunt and John Shelley, *Computers and Commonsense,* Prentice-Hall, London (1975) p. 125

57 T. F. Fry, *Computer Appreciation,* Butterworths, London (1970) p. 8

58 Roger Hunt and John Shelley, *Computers and Commonsense,* pp. 127–9

59 T. F. Fry, *Computer Appreciation,* pp. 9–10

60 Harold Laswell, B. J. Schafer and Cuthbert Hurd, 'Social problems of automation', *Proceedings of the Western Joint Computer Conference,* American Institute of Electrical Engineers (March 1959) pp. 7–14

61 Henry Jacobowitz, *Electronic Computers Made Simple,* pp. 1–4

62 Edward Tomeski, *The Computer Revolution,* Macmillan, New York (1970) pp. 53–62

63 S. H. Hollingdale and G. C. Toothill, *Electronic Computers,* p. 316

64 Roger Hunt and John Shelley, *Computers and Commonsense,* pp. 78–90. See also Kenneth Owen, ' Scanning new vistas in medical electronics', *The Times* (6 October 1976); 'Automation could make libraries obsolete' *The Times* (26 February 1977) p. 3

65 Murray Laver, *Introducing Computers,* H.M.S.O., London (1973) pp. 58–9; William Desmonde, *Computers and Their Uses,* Prentice–Hall, New Jersey (1971) pp. 4–10

66 Parsons and Williams, *Forecast 1968–2000 of Computer Developments and Applications,* Copenhagen (1968)

67 Eric Ashby, 'A second look at doom', *Esso Magazine* (Spring 1976) 3

INDEX